NANOTECHNOLOGY SCIENCE AND TECHNOLOGY

GOLD NANOPARTICLES

SYNTHESIS, OPTICAL PROPERTIES AND APPLICATIONS FOR CANCER TREATMENT

NANOTECHNOLOGY SCIENCE AND TECHNOLOGY

Additional books in this series can be found on Nova's website under the Series tab.

Additional E-books in this series can be found on Nova's website under the E-book tab.

MATERIALS SCIENCE AND TECHNOLOGIES

Additional books in this series can be found on Nova's website under the Series tab.

Additional E-books in this series can be found on Nova's website under the E-book tab.

NANOTECHNOLOGY SCIENCE AND TECHNOLOGY

GOLD NANOPARTICLES

SYNTHESIS, OPTICAL PROPERTIES AND APPLICATIONS FOR CANCER TREATMENT

ABAGAIL JARNAGIN
AND
LOGAN HALSHAUSER
EDITORS

nova publishers
New York

For permission to use material from this book please contact us:
Telephone 631-231-7269; Fax 631-231-8175
Web Site: http://www.novapublishers.com

Library of Congress Cataloging-in-Publication Data

Gold nanoparticles : synthesis, optical properties and applications for cancer treatment / editors, Abagail Jarnagin and Logan Halshauser.
p. cm.
Includes index.
ISBN 978-1-62257-927-3 (hardcover)
1. Nanoparticles. 2. Nanoparticles--Therapeutic use. 3. Gold--Therapeutic use. I. Jarnagin, Abagail. II. Halshauser, Logan.
RS201.N35G65 2013
615.2656--dc23

2012034866

Published by Nova Science Publishers, Inc. † New York

CONTENTS

PREFACE

In this book, the authors present current research in the study of the synthesis, optical properties and applications for cancer treatment of gold nanoparticles. Topics discussed include the use of gold nanoparticles in cancer treatment and biomedical applications to target tumors and provide detection, drug carriers, gene silencing and radiotherapy; gold nanoparticle fabrication by laser ablation technique and their optical and morphological study; gold nanoparticles for metabolite imaging; formation of gold nanoparticles inside the corona of amphiphilic triblock copolymer micelles; and the intracellular delivery of gold nanoparticles and their application in nanomedicine.

Chapter 1 – The different methods for preparing gold nanoparticles will be referred, namely the traditional deposition-precipitation and co-precipitation techniques, variations of impregnation, reductive deposition, vapour-phase methods, ion exchange and colloidal gold. Also a brief history of the use of gold in medicine throughout the ages will be told. More recent uses of gold in cancer treatment and (bio)medical applications, to target tumors and provide detection, drug carriers, gene silencing, radiotherapy, etc, will also be referred.

Chapter 2 – Material particles of the size of few nano-meters to couple of hundreds of nano-meters exhibit size and shape dependent properties and the consequent versatility and importance in their use have made them an exciting and interesting field of research. Of these, noble metal nano-particles, particularly gold nano-particles have shown tremendous potential for their wide range of applicability not only in fundamental sciences like physics, chemistry and biology, but also in applied sciences such as in designing and developing varieties of nano-particle based devices and sensors. The number of articles published on noble metal nano-particles, particularly gold and silver nano-particles has gone up in exponential rate since the early nineties of the last century.

One of the major thrust areas of nano-particles and their science and technology research are their fabrication. In the nineties of the last century and the first half of the last decade chemical methods dominated over all other methods, including the laser ablation technique which started emerging. The main disadvantage including a host of other drawbacks of conventional nano-particle generation techniques such as mechanical milling and grinding or even the chemical method like the sol-gel process is that it almost invariably picks up impurities which make them unusable and hence undesirable for many applications e.g. biological applications.

Pulsed laser ablation is an attractive and a promising alternative, addressing the shortcomings of the conventional methods, including the chemical methods.

In this chapter the authors discuss mostly about the nano-particle fabrication, with particular emphasis on gold nano-particles in colloids, using nano second laser pulses. A number of variables of the ablating laser pulse have been used to control the size of the fabricated nano-particles. These were done without the use of chemicals/chemical surfactants, which provided a good and clear understanding of the laser parameters on the control of the nano-particle size. Possible explanations of all observations are given.

Chapter 3 – Gold nanoparticles have been drawing a lot of attention because of their novel physicochemical properties, especially their special nonlinear optical properties and surface-enhanced Raman scattering effect, and potential applications in biology, catalysis, photocatalysis, and microelectronic.. High affinity of organic functional groups like –SH, -CN, -COOH, $-NH_2$ for metallic surfaces such as Au, Cu, Ag, Pd, Pt, Hg have motivated several research groups to focus on using this phenomena for industrial applications. The self assembling of some synthesized thiol surfactants namely (mercaptopropane-, mercaptohexane-, mercaptooctane- and mercaptodecane sodium sulfonate) on the fabricated gold nanoparticles was investigated. The self assembling of these surfactants on gold nanoparticles was characterized using different techniques. The effect of self-assembling of these surfactants on the size of the gold nanoparticles was studied using TEM images. The growth of the gold nanoparticles was investigated with respect to the increase of alkyl chain in the synthesized thiol surfactants (Azzam and Badawi et al 2009). The fabrication of surface plasmon resonance (SPR) biosensors using self-assembled monolayers (SAMs) of 1,10-decanedithiol surfactant and adsorbed gold nanoparticles was investigated. Several spectroscopic and microscopic techniques were used to investigate both the SAM and the chemisorption of gold nanoparticles at the SAM surface. Surface plasmon resonance (SPR) was used to study both the adsorption of Bovine serum albumin (BSA) as tested protein to the SAM surface and to the gold nanoparticle-coated SAM (Azzam and Grunwald et al 2009). The self assembling behavior of the synthesized dodecylcysteine hydrochloride surfactant comparing with that of cysteine compound on the prepared gold nanoparticles was confirmed using transmission electron microscope (TEM) measurements. The antitumour activity of the prepared surfactant without and with the gold nanoparticles was investigated (Azzam et al 2008).

Chapter 4 – Gold nanoparticles (Au NPs) have attracted significant attention due to their unique optical properties which differ from those observed in the bulk material. Their size, shape and morphology, as well as the surrounding medium are the key factors that enhance and tune these properties, allowing for the application of AuNPs in photonic devises, biosensors, nanoreactors and in biomedicinal applications. Usually, amphiphilic block copolymer micelles are used for the formation of AuNPs, either inside their cores or into their corona, depending on the envisioned application. In this work, poly[(methacrylic acid)-b-(styrene)-b-(methacrylic acid)] (PMAA-PS-PMAA) triblock copolymer was utilized for the *in situ* synthesis of Au NPs inside the corona of the copolymer micelles in aqueous media. Stoichiometric equivalents of $HAuCl_4$ were added as the metal precursor, which were preferentially dissolved and coordinated to the coronas of the block copolymer micelles and were subsequently reduced to Au NPs without addition of any reducing agent. Therefore, the block copolymer acts as the nanotemplate and the carboxylate groups as the reducing agents. UV-vis absorption spectroscopy was employed in order to follow the Au NPs formation kinetics. Dynamic light scattering (DLS) was utilized for studying the colloidal stability and structural response of PMAA-PS-PMAA/Au NPs hybrid nanostructures, as a function of the pH, the ionic strength and the temperature of the solution. Additional information on the structure of the nanocolloidal system, the polymer-inorganic material interactions and metal nanoparticles characteristics were obtained by the use of atomic force microscopy (AFM) and transmission electron microscopy (TEM).

Chapter 5 – Matrix-assisted laser desorption/ionization (MALDI)-imaging mass spectrometry (IMS) enables visualization of the distribution of metabolites in tissue sections. IMS is a powerful technique for the visualization of endogenous small metabolites, especially lipids, facilitated by the advantages of mass spectrometry-based molecular detection. Despite the promising capability of MALDI-IMS for imaging, there is still some room for improving the technique, especially in spatial resolution and sensitivity.

In this review, the authors show some applications of gold nanoparticles (AuNPs) for mapping the distribution and localization of small metabolites in tissue sections. The authors have developed and applied AuNPs as a new matrix to maximize the detection of metabolites in the mouse brain. The authors believe that the authors' approach using AuNPs in IMS could lead to a new strategy for analyzing basic biological mechanisms and several diseases.

Chapter 6 – Nanotechnology has enormous potential in biomedical research including disease diagnosis and therapy. Inorganic nanoparticles have been adopted for biomedical research due to their unique optical and physical properties. Compared to conventional materials, inorganic nanomaterials have several advantages such as simple preparative process and precise control over their shape, composition and size. A major aim of medicine is the early and precise diagnosis of clinical conditions, providing an efficient treatment without secondary effects. Nanoparticle-based technologies have demonstrated especially high potential for medical purposes, ranging from diagnosing diseases to providing novel therapies. Among the metal nanoparticles, gold nanoparticles have proven to be powerful tools in various nanomedicinal applications because of their photo-optical properties and biocompatibility. However, the existing nanoparticle-based technologies must overcome several challenges, including selective nanoparticle delivery, potential toxicity, imaging of nanoparticles and therapeutic efficacy. This chapter describes a brief overview of the recent research on delivery of gold nanoparticles inside the cells and their potential applications in nanomedicine.

Chapter 7 – The authors report synthesis of water-soluble gold nanoparticles (Au-NPs), having an average diameter of ca. 20 nm, by reducing AuCl3 using ciprofloxacin (CF) as a reducing/stabilizing agent. The synthesized Au-NPs have been characterized by scanning electron microscopy (SEM), EDS, TEM, UV-visible spectroscopy, X-ray diffraction and cyclic voltammetry. TEM and SEM combined with EDS confirmed that spherical shaped Au-NPs were formed. Uv-vis spectra of Au-NPs show two absorption bands corresponding to the capping agent ciprofloxacin and surface plasmon absorption bands at 274 nm and 535 nm, respectively. The Au-NPs are used to modify a glassy carbon electrode (GCE) and it's electrochemical and electrocatalytic properties are investigated. The Au-NPs show excellent electrocatalytic activity toward the oxidation of methanol at +0.33 V in alkaline solution, which was not observed on unmodified GCE. Further, Au-NPs modified electrode showed excellent electrocatalytic properties for reduction of oxygen at lower potential. Here, CF was used as a reducing agent to synthesis of Au-NPs. The synthesized Au-NPs are highly stable, can be stored for more than three months in air at room temperatures.

In: Gold Nanoparticles ISBN 978-1-62257-927-3
Editors: A. Jarnagin and L. Halshauser

Chapter 1

SYNTHESIS AND APPLICATIONS OF GOLD NANOPARTICLES

Sónia Alexandra Correia Carabineiro
Laboratório de Catálise e Materiais (LCM), Laboratório Associado LSRE/LCM, Faculdade de Engenharia, Universidade do Porto, Portugal

ABSTRACT

The different methods for preparing gold nanoparticles will be referred, namely the traditional deposition-precipitation and co-precipitation techniques, variations of impregnation, reductive deposition, vapour-phase methods, ion exchange and colloidal gold. Also a brief history of the use of gold in medicine throughout the ages will be told. More recent uses of gold in cancer treatment and (bio)medical applications, to target tumors and provide detection, drug carriers, gene silencing, radiotherapy, etc, will also be referred.

INTRODUCTION

For many centuries gold was merely a beautiful precious glittering metal, only in the mind of coiners and inspired craftsmen who shaped it into beautiful ornamental pieces of art, which were also symbols of power and wealth. Therefore, for many centuries this metal was out of the mind of scientists, due to the prior experience with platinum group metals that exhibit both a high chemisorption power and a high catalytic activity in several reactions. It was then believed that the rather low chemisorption ability of gold implicated a low catalytic power as well.

However, in recent years, it began to be realized that, when applied to supports as nanoparticles, gold becomes really active, and so the research in the field increased enormously [1-11], and nowadays there a great expectation about the catalytic potential of this metal for several reactions. Gold particles are readily prepared on the nanoscale and are being used in this form in a wide range of practical applications, including the biomedical monitoring of constituents in body fluids, and in decorative inks, cosmetics, and lubricants, and there is a rapidly growing interest in their use as catalysts. For gold to be active, careful preparation procedures are crucial in order to obtain it as well dispersed nanoparticles. This chapter focus on its preparation methods and applications.

Preparation Methods

Colloidal Gold

Colloidal gold nanoparticles can be usually found in liquid suspensions (usually in water). They are permanently suspended in solution and have properties different than supported gold nanoparticles or bulk gold. Different colours can be found according to the size and shape of the nanoparticles in the solution (Figure 1). Particles with a diameter around 10 nm are ruby red due to a narrow absorption band at 520 nm [10, 12-14]. Particles smaller than 40 nm show negligible scattering of light. As particle size increases, the absorption band shifts to larger wavelengths and light scattering is observed. If particles coagulate at around 100 nm, or show a rod-like form, then purple, violet or pale blue colours are found and light scattering accompanies absorption [10, 12-14]. Naturally, bulk gold is bright yellow, and the variation of the colours of colloidal gold nanoparticles with their size depends on light interaction with matter [14].

Colloidal gold has a very long history. Colourful aqueous solutions of gold colloids date back to Roman times and were known to the medieval alchemists [15]. The notion that gold sols contain small metallic particles was first noted by Michael Faraday as early as 1857 [11], who used several agents to reduce a gold chloride solution and produce gold sols. New recent developments in preparative and analytical methods have led to the preparation of much smaller particles and to the determination of intimate details of their structures [10, 14, 16].

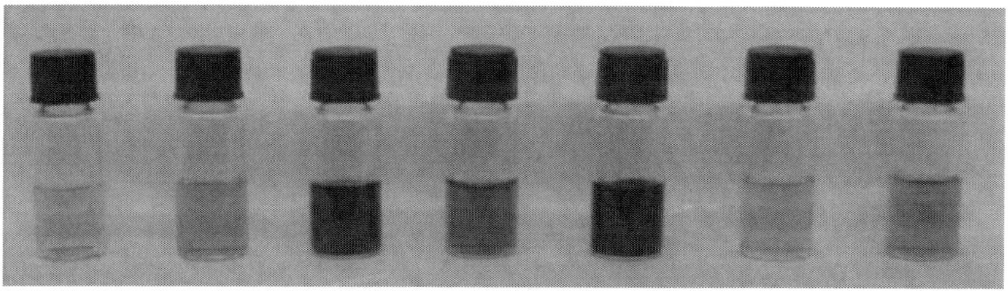

Figure 1. Colloidal gold nanoparticles showing different colours in solution.

Two basic approaches are used in the preparation of colloidal gold solutions. One is the disintegration of metallic gold rods by an electric arc operating in a liquid medium, used by Georg Bredig [17]. An arc is struck between electrodes (Figure 2), under the surface of water containing a stabilizing agent (traces of KOH, for example). The intense heat of the arc vaporizes some of the metal, which then condensed under cold water, kept in an ice bath. The colloidal gold particles are then produced by the fast condensation of the vapour.

Theodor Svedberg pioneered the use of electrochemical methods for the synthesis of gold particles, modifying Bredig's method [18, 19]. Just like Faraday, Svedberg reported the use of almost every conceivable reducing agent available at his time, in order to produce colloidal gold from hydrochloroauric acid [18].

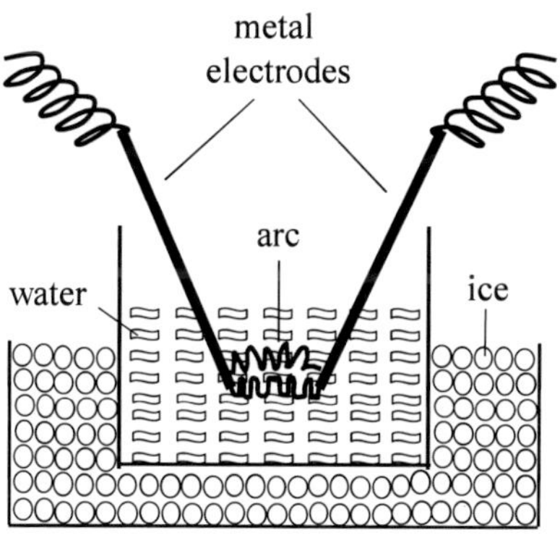

Figure 2. Bredig's arc set-up for producing colloidal gold (adapted from [15]).

The other approach is reduction of chloroauric acid or other gold salts by radiation or several reducing agents [10, 17]. This reduces the Au^{3+} ions to neutral Au atoms. As more gold atoms are formed, the solution becomes supersaturated, and gold gradually starts to precipitate in the form of nanoparticles. A stabilizing agent is usually added to prevent the particles from aggregating. The preparation can be influenced by several factors such as temperature and reactant concentrations [10, 20]. The product is a hydrosol of gold of a specific colour, depending on particle size (Figure 2). Control over the gold colloidal particle size, by simple variation of the relative surfactant concentration in the salt solution during reduction has been studied in detail for alkanethiols by Leff and co-workers [21].

One of the common methods for producing colloidal gold is the method developed by Turkevich [22, 23], which was refined by Frens [24, 25]. It is simple and commonly used to produce spherical gold nanoparticles of around 10 nm in diameter. It involves the reaction of small amounts of hot chloroauric acid with small amounts of sodium citrate solution. Colloidal gold is formed because the citrate ions act both as reducing and capping agents.

Brust et al. [26] developed a colloidal gold technique that allowed to produce gold nanoparticles in organic liquids that were normally not miscible with water (like toluene). It involves the reaction of a chloroauric acid solution with tetraoctylammonium bromide (TOAB) solution in toluene (acting as an anti-coagulant, transfer catalyst and stabilizing agent) and sodium borohydride ($NaBH_4$, acting as a reducing agent). The gold nanoparticles obtained were around 5–6 nm [26].

Stable “naked” gold nanoparticles in water can also be obtained through reduction of $HAuCl_4$ with $NaBH_4$ without addition of a stabilising agent, with possibility to tune the sizes from 3.2 to 5.2 nm [27]. This synthesis method is simple, cheap, greener, can have for many practical applications by combining the easiness of the Turkevich method with the monolayer-protection of the Brust method. The advantage is that no non-volatile and difficult to remove phase transfer agent, such as TOAB, is required.

Perrault and Chan’s [28] reported the use of hydroquinone to reduce aqueous $HAuCl_4$. This seed-based synthesis method is related to that used in photographic film development, in which the growth of silver grains within the film occurs after addition of reduced silver onto the surface. In a similar way, gold nanoparticles, along with hydroquinone, catalyse reduction of ionic gold onto their surface. Addition of citrate as a stabilizer allows a controlled particle growth. Usually, nanoparticle seeds are produced using the citrate method. The hydroquinone method complements that of Frens [24, 25], as monodispersed spherical particle sizes can be obtained. Frens method produces particles of 12-20 nm, while Perrault's method originates particles of 30-250 nm.

Sonolysis can also be used for the synthesis of colloidal gold. This procedure is based on ultrasound and reduction of an aqueous solution of $HAuCl_4$. Au-NPs with multiple shapes (spherical, rod, triangular plate and disk) and size (5-50 nm) were synthesized by controlling the concentration of sodium dodecylsulfate and ultrasonic irradiation power [29].

Ultra-violet irradiation can also be used. Gold nanoparticles and nanoplates with obtuse triangle, hexagon and round shapes with sizes from 2 to 10 nm were synthesized this way in the presence of citric acid and poly(vinylpyrrolidone) at room temperature [30].

Supported Gold Nanoparticles

There are several techniques described in the literature to obtain gold well dispersed on supports. The most common are referred below.

Impregnation

This is the classical method used to prepare supported metal catalysts and consists of simply impregnating a support with a solution of a metal salt. This usually involves suspending the support in a larger volume of solution, from which the solvent is then removed. An alternative variation is the so-called incipient wetness (IW) technique, in which the pores of the support are filled with the solution.

Common gold precursors like chloroauric acid ($HAuCl_4$) or auric chloride ($AuCl_3$ or Au_2Cl_6) are usually used but complex salts, such as potassium aurocyanide ($KAu(CN)_2$) and the ethylenediamine complex $[Au(en)_2]Cl_3$, are also alternatives. Traditional supports are silica, alumina and magnesia, but titania, alumina, boehmite (AlO(OH)), ferric oxide (α–Fe_2O_3) and magnesium hydroxide can also be used [2]. After drying, the precursor needs calcination at temperatures as high as 1073 K. Reduction by hydrogen at 523 K, aqueous oxalic acid at 313 K or aqueous magnesium citrate [2] is also needed.

Conventional impregnation techniques are not so good for gold catalysts as they result in large gold particles, which are more likely to be inactive [2, 10, 11, 31, 32]. An example is shown in Figure 3a. Morevoer, it is difficult to obtain a high dispersion of gold. Also, during calcination, the gold particles experience severe agglomeration as $HAuCl_4$ interacts only weakly with the support [1, 10, 31, 33]. The presence of chloride also promotes sintering [10, 11, 32]. The early use of classical impregnation techniques was the main reason why catalysis by gold was thought for so long to be inactive, in comparison with other noble metals such as Pt and Pd.

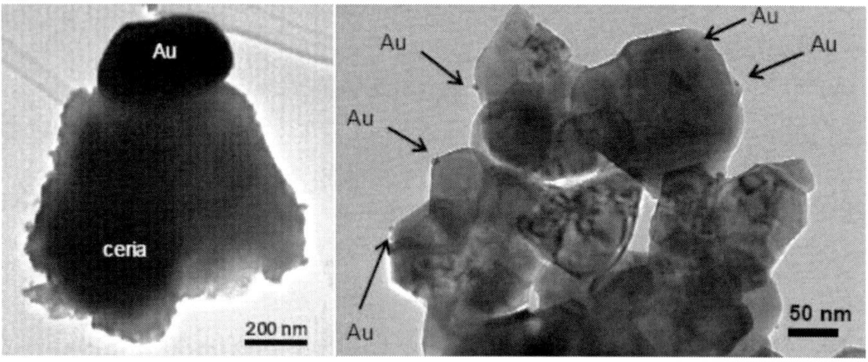

Figure 3. TEM images of Au/CeO_2 prepared by a) traditional impregnation and b) double impregnation (adapted from [32, 34]).

However, Datye and co-workers reported an alternative impregnation method for supporting gold on alumina using $HAuCl_4$ [35]. Since impregnation under acidic conditions leads to poor dispersion of Au and the resulting catalysts are not very active, a two-step procedure has been developed: In the first step, gold chloride was adsorbed on alumina from an acidified solution. After washing off the excess gold precursor, the solid was treated with strong base to convert the chloride to an adsorbed hydroxide. Drying and calcining at 673 K yielded a catalyst with gold particles having an average diameter of 2.4 nm, with good activity, and stability to hydrothermal sintering. This new method is a successful impregnation preparative route for gold catalysts, allowing them to have durability at least until 873 K, since no chloride is present.

Also Bowker et al. refer the use of a double impregnation method that removes chloride [36]. Briefly, this method consists in impregnating the support with an aqueous solution of the gold precursor ($HAuCl_4$) and then with a solution of Na_2CO_3, under constant ultrasonic stirring, followed by washing with water and drying in an oven overnight at 393 K. These authors used this method to prepare Au/TiO_2 materials, and state that it represents an environmentally and economically favourable route to high activity gold catalysts production. This method has been used with great success with other supports as well [34, 37-42]. One example is shown in Figure 3b.

Co-Precipitation

This is one of the simplest ways to prepare gold catalysts and it was one of the first to be used [9, 10]. An aqueous solution of $HAuCl_4$ and water-soluble metal salts, such as a nitrate, is poured into an aqueous alkaline solution (Na_2CO_3 and/or NH_4OH) and stirred for a few minutes. The two hydroxides (or hydrated oxides) are then precipitated simultaneously. After ageing for about 1 h, the precipitates are washed several times with water and filtered. The hydroxide and/or carbonate mixture is dried overnight and calcined to obtain powder catalysts [10, 43, 44]. Then a reduction step is necessary. It is possible that some catalysts prepared by this method contain a significant concentration of Na^+ and Cl^- ions, if a metal chloride is used as precursor. Both can act as a catalyst poison [1, 2, 10]. The applicability of this technique is limited to metal hydroxides or carbonates that can be co-precipitated with $Au(OH)_3$.

Deposition Precipitation

This method is convenient and can be used for producing commercial gold supported catalysts [6, 10] and applied to the widest range of different support materials [45]. The precursor to the active species is brought out of solution in the presence of a suspension of the support, usually by raising the pH in order to precipitate a hydroxide. The surface of the support acts as a nucleating agent, and the method leads to the greater part of the active precursor being attached to the support. After the pH of an aqueous solution of $HAuCl_4$ is adjusted with NaOH, to a fixed point in the range of 6–10, a metal oxide support, can be immersed in the solution. The partially hydrolysed species $[Au(OH)_nCl_{4-n}]^-$ (n = 1–3) then react with the surface of the support. Ageing for about 1 h results in the deposition of $Au(OH)_3$ exclusively on the surface of the metal oxide support, if the concentration and temperature are properly chosen [1, 2, 10].

The influence of the pH on the particle size of Au is remarkable, as above pH 6, the main species of Au in solution are transformed from $AuCl_4^-$ to $[Au(OH)_nCl_{4-n}]^-$ (n = 1–3), and the mean particle diameters of Au in the calcined catalysts become smaller than 4 nm [1]. Several authors claim that a pH ranging from 7 to 8 is preferable depending on the oxide support [45-47], since at this pH, the value of *n* is close to 3, and at lower values of pH, there is less hydrolysis of the Au–Cl bond. However, Bond and co-workers [48] showed that pH 9 was the optimum value to be reached during deposition precipitation for Au/TiO_2 catalysts. At that pH, the main species in solution are anionic Au complexes, with almost no chloride, while at lower values, the Au complexes contained chloride, Au particles were larger, and activities lower.

An alternative method for adjusting the pH is to use urea [2, 10, 49-52]. This consists in the slow decomposition of urea in the solution. Hydroxyl ions are generated slowly and uniformly throughout the liquid phase, and their concentration is always low because they are consumed almost as soon as they are formed. Louis and co-workers found that the use of urea yielded the same gold particle sizes as those obtained using NaOH (2–3 nm), and no sodium poison was introduced [49].

A constraint of the deposition-precipitation is that it is not applicable to activated carbon [6, 10, 11, 53] or zeolites [54], due to their high isoelectric point. However, it has the advantage over co-precipitation in that all of the active component remains on the surface of the support and none is buried within it [2, 10, 46]. Also, it gives narrower particle size distributions, but it is recommended that the support should have a surface area of at least 50 m^2/g [2, 10].

Ion-Exchange

In this method ions on the surface of the support are replaced by gold ions. The procedure is especially effective with zeolites. But the introduction of catalytically active species into the cavities of these materials, as opposed to placing them on their external surface, presents certain difficulties, namely the lack of suitable cations or cationic complexes [2]. Nevertheless, different kinds of Au/zeolite systems have been prepared this way [55-60].

Pitchon and co-workers have developed a novel method for preparing supported gold catalysts based on the direct anionic exchange (DAE) of the gold species with the hydroxyl groups of the support [50, 61-65]. An aqueous solution of $HAuCl_4$ is added to the support, heated to 343 K and aged for 1 h. The slurry is then filtered, washed with warm water, dried in an oven at 393 K overnight and calcined in air at 573 K for 4 h. In order to completely remove the chloride ions, these authors take a fraction of the catalyst after drying and wash it with a concentrated solution of ammonia. However, this may be a dangerous procedure, since gold oxide and ammonia sometimes produce fulminating gold, which is explosive [66, 67].

Liquid-Phase Reductive Deposition

This method consists of mixing a solution of the gold precursor ($HAuCl_4$) with a solution of NaOH with stirring at room temperature. The resulting solution is aged for 24 h, in the dark, at room temperature to complete the hydroxylation. Then the appropriate amount of support is added to the solution and, after ultrasonic dispersion for 30 min, the suspension is aged in the oven at 373 K overnight. The resulting solid is washed repeatedly with distilled water for chloride removal, and dried in the oven at 373 K overnight.

The selective reductive deposition is performed by adsorption of the gold ions onto the surfaces where the reduction takes place. Thus, the initial adsorption is the key feature of this technique, and the key points are precise control of the metal complex by adjusting solute conditions, such as composition and structure of the metal complex; storing of the suspension until the equilibrium composition is attained, and aging the suspension at a controlled temperature. Using this new method, Au nanoparticles supported on various carriers were successfully obtained [32, 41, 42, 68]. An example is shown in Figure 4.

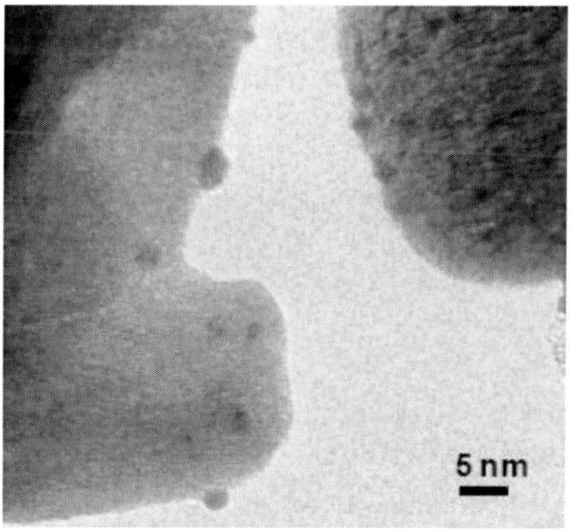

Figure 4. TEM image of Au/La_2O_3 prepared by and liquid-phase reductive deposition (adapted from [41]).

Vapour-Phase Methods and Grafting

These methods are similar, differing only if a solvent is present or not. In the vapour-phase method (*Chemical Vapour Deposition*) a stream of a volatile compound of gold is transported onto a high area support by an inert gas and it reacts chemically with the surface of the support to form a precursor to the active species [1, 2, 10, 11]. The most widely used gold precursor is $AuCl_3$ or $HAuCl_4$, but other substances have also been used, mainly to prevent chloride contamination.

In *Physical Vapor Deposition*, gold is vaporized from a target under vacuum and deposited on an oxide support or carbon under high vacuum conditions [11].

The 3M company in Minnesota (USA) has found that very active gold nanocatalysts can be prepared this way on a wide range of supports, including some that are water soluble, or not suitable for deposition-precipitation, like SiO_2 [69, 70].

This method is low-cost, has great reproducibility, does not need washing and thermal treatment steps, as do those materials resulting from solution preparation methods, and has no toxicity hazards.

In the *grafting* method, a gold complex in solution reacts with the surface of a support, forming species convertible to a catalytically active form.

Thus, gold phosphine complexes have been grafted onto the surface for a number of freshly precipitated wet hydroxides [2, 10, 71-74], since they have many surface –OH groups reactive to the Au compounds.

Vacuum-drying at room temperature and temperature-programmed calcination in air follows, causing simultaneous transformation of both precursors to gold particles and oxides, respectively, under their chemical interactions by temperature-programmed calcination.

Au–phosphine complexes are choice candidates for metal precursors because they thermally decompose to Au metal in a temperature range similar to that used for the transformation of wet metal hydroxides to oxides. Moreover, the phosphine ligands are expected to retard the growth to large Au metallic particles.

Gold can be deposited on SiO_2, MCM–41, SiO_2–Al_2O_3, activated carbon as nanoparticles with high dispersion by the gas-phase grafting of an acetylacetonate complex of gold, while liquid-phase preparation methods are not that effective with these supports [75].

USES AND APPLICATIONS OF GOLD

Use as a Catalyst

Gold nanoparticles on several supports have been used as catalysts for several reactions, namely for the *oxidation of carbon monoxide* ($CO + \frac{1}{2} O_2 \rightarrow CO_2$) at ambient temperatures and below, as first described by Haruta [1, 3-7]. This reaction is important for air-purification devices for indoor space, gas masks, cleaning automotive exhaust gases and sensors [2, 9-11, 15, 32, 34, 37-39, 41, 42, 76, 77]. It is particularly advantageous that gold's catalytic activity is often promoted by moisture [78, 79]. In fact, water can increase the rate by two orders of magnitude. Prototype products that use gold catalysts for low-temperature air quality control are now appearing in the public domain [9-11].

Gold catalysts are also active for the oxidation of CO in an H_2 rich environment, also called *preferential oxidation (PROX)*, which has considerable technical interest for purification of hydrogen feed gas to use as a supply in fuel cell technology [9-11, 80-88]. The advantage of gold over other metals like platinum is that it oxidises CO (that acts as a poison for fuel cells) to CO_2 (that is harmless to such devices) and not H_2 (while other catalysts oxidise both CO and H_2).

The *water gas shift* reaction ($CO + H_2O \rightarrow CO_2 + H_2$) is of interest, both for the production of pure hydrogen for fuel cells and heavy chemical processing and is one of the key steps involved in automobile exhaust processes. Gold nanoparticles have been used with very good results [9-11, 89-95].

Selective oxidation is used in the activation of raw materials to provide useful products and chemical intermediates. Gold nanoparticle based catalysts may be used under mild conditions with gaseous oxygen or air in aqueous dispersion to promote these transformations effectively. The oxidation of aldehydes, alcohols, amines, diols, glycerol, glucose, hydrocarbons, ammonia, propene and styrene oxide, etc, has been achieved with great success. [9-11, 96-111].

Selective hydrogenation of several organics such as α,β-unsaturated aldehydes, α, β-unsaturated ketones and unsaturated hydrocarbons [9-11, 112-126]. These reactions of considerable interest in the industrial synthesis of fine chemicals, particularly, of pharmaceuticals and cosmetics

Reduction of nitrogen oxides (collectively known as NO_x) to N_2 with propene, carbon monoxide or hydrogen has also been achieved with gold catalysts [9-11, 127-131]. These reactions are of extreme importance since

NO_x are emitted into the air during the combustion of fossil fuels and can dissolve in water to form nitrous and nitric acids which cause acid rain or combined with hydrocarbons (also emitted during fossil fuel combustion) to produce photochemical smog in cities.

Gold nanoparticles have also been used in the *decomposition of dioxins*, that are heterocyclic hydrocarbons, mostly poisonous chemical by-products of the manufacture of certain herbicides and bactericides [9-11, 132] and in the *oxidative decomposition of volatile organic compounds* [9-11, 133-145], that a major contributor to air pollution because of their toxic and malodorous nature and their contribution to ozone and smog formation.

Examples of other reaction catalysed by gold nanoparticles with emerging commercial applications can be found in several reviews on the subject [1-7, 9-11, 76, 77].

Use in Medicine throughout the Ages

Gold was used for medical purposes already in ancient Egypt [15, 146, 147]. There, some alchemists worked on the development of a colloidal elixir known as "liquid gold," which supposedly had the ability to restore health and youth. In fact, the Bible tells that Moses, after delivering the children of Israel from slavery in Egypt, came down from Mount Sinai with the tablets of the Ten Commandments given by God, and found out that the people had taken their gold jewellery and had made an image of a golden calf and were worshiping it, as they got tired of waiting for Moses that had gone up to the mountain. In Exodus 32:19-20 it can be read:

> "And it came to pass, as soon as he came nigh unto the camp, that he saw the calf, and the dancing: and Moses' anger waxed hot, and he cast the tables out of his hands, and brake them beneath the mount (see Figure 5), and he took the [golden] calf which they had made, and burnt it in the fire, and ground it to powder, and stowed it upon the water, and made the children of Israel drink of it."

This might be considered the first documented mental and spiritual health recipe given in the history of mankind for a crude "colloidal gold" formula, prescribed by Moses for the impatience and disobedience of the people of Israel.

Figure 5. "Moses breaks the Tablets of Stone" (image used with permission from La Vista Church of Christ).

Also the ancient Chinese healers used gold solutions, although their methods were lost meanwhile. The hindu Ayurveda, the ancient system of Indian medicine, also refers to the use of colloidal gold, prepared by the reduction of gold salts with several organic substances [148]. In ancient Rome, gold was used for the treatment of skin lesions, ulcers fistulas, haemorrhoids and warts [15, 146].

The medieval alchemists also used gold for in their elixirs for the health and life extension. The so-called "Elixir of Life", a potion made from gold, was mentioned already in ancient times [15, 149]. In fact, Paracelsus (1493-1541) made a colloidal gold solution, to which he called "Aurum Potabile" (drinkable gold in Latin) [146, 147, 149]. In Europe, in medieval times, colloidal gold solutions ("gold waters") were extremely popular [15, 146].

In the 17th century, gold began entering the official drug pharmacopoeias of England and Germany, who accepted several different gold preparations [15, 146, 149]. Alchemists mixed gold in powder into drinks to treat sore limbs. In fact, in modern medicine gold is still used to treat arthritis, as it will be discussed ahead. In the late 19th century, several gold preparations were used in the treatment of syphilis, goitre and scrofula [150]. These results influenced the medicine of Europe and United States, where it was used in the treatment of syphilis [151], alcoholism [152], for the healing activity of the heart and improved blood circulation, skin ulcers, burns, etc [15, 146].

Already in the 20th century, it was reported that beneficial results were obtained with colloidal gold for cancer disease [153]. Nowadays, the gold is mostly used in the treatment of rheumatoid arthritis [15, 146]. Gold-198 (a radioisotope of gold emitting gamma and beta radiation) is indicated to intracavitary administration as it is chemically inert, inhibits the formation of cavity fluid and has a relatively short half-life of 2.7 days [154]. Au-198 has been used also in the treatment of persistent knee effusions [155, 156] and also rheumatoid arthritis [157-162].

Metallic gold is approved as a food additive in the EU (E175 in the Codex Alimentarius), due to its relative chemical inertness, and resistance to corrosion or transformation into soluble salts. In fact, pure metallic (elemental) gold is non-toxic and non-irritating when ingested, and is sometimes used as a food decoration in the form of gold leaf and some drinks contain also gold [149]. However, soluble compounds, like gold salts, namely gold chloride and gold cyanide, can be toxic.

Presently, gold can become useful for applications to "lab on a chip" devices, electronic, photonic and sensing applications based on plasmonics, cancer treatment, etc [14, 15, 163-166]. More details are given below.

Treatment of Rheumatoid Arthritis

Gold nanoparticles have well known anti-inflammatory properties and have been successfully used to treat rheumatoid arthritis [155-162, 167-176], which is a autoimmune, chronic inflammatory disorder that affects many tissues and organs, but specially joints (Figure 6), leading to the destruction of articular cartilage and ankylosis of the joints. The anti-inflammatory activity of colloidal gold can be greatly increased by using the liposomes to target the inflamed joints, showing also the beneficial effect of reduced cartilage destruction [177]. Also the implantation of gold beads near arthritic hip joints has been found to relieve pain in dogs [178].

The latest gold rush is likely to revolutionize the field of biosensing and chemical sensing, by allowing to develop new technologies that can help identifying chemical or biological strains with high accuracy, by using highly environmentally sensitive nature of plasmon resonance as well as other optical effects (such as surface-enhanced Raman or fluorescence) associated with gold and other noble metal particles [14, 15].

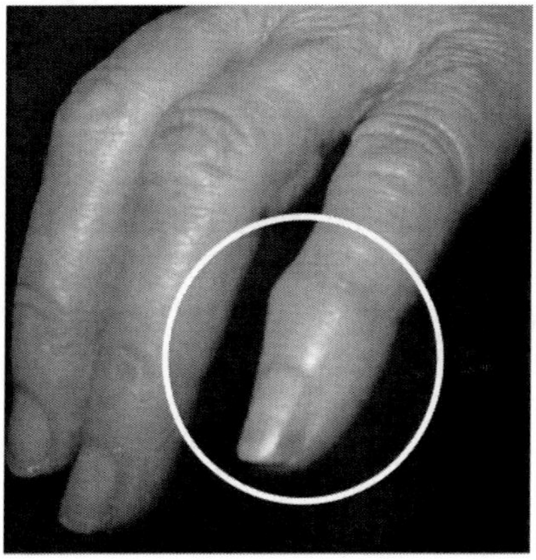

Figure 6. Swelling and stiffness in a finger joint caused by arthritis.

Use as Contrast Agents

Gold nanoparticles can be used as contrast agents in biological electron microscopy or computed tomography [179-188]. They can be attached to many traditional biological probes such as antibodies, lectins, superantigens, glycans, proteins, nucleic acids, and receptors [189, 190]. Particles of different sizes are also distinguishable in electron micrographs, allowing simultaneous multiple-labelling experiments [191, 192].

Tumor Targetting

Gold nanoparticles can be used in cancer research to target tumors and provide detection [166, 193-201]. PEGylated (polyethylene glycol coated) gold particles are conjugated with an antibody, which is sometimes overexpressed in cells of certain cancer types to specifically target tumor cells. These PEGylated gold nanoparticles can then detect the location of the tumor using Surface Enhanced Raman Spectroscopy [195]. PEGylated gold nanoparticles also strongly accumulate in tumors in mice [201]. It was shown that 20 nm Au nanoparticles coated with thioctic acid-terminated PEG are promising potential drug delivery vehicles and diagnostic imaging agents [197]. Besides PEG, other functionalising agents such as cyclodextrin [198], gellan gum [202], chitosan [203], aspartic acid [204], glutamic acid [186], etc, can also be used.

Drug Carriers

Clinically used drugs are usually low molecular-weight compounds that diffuse rapidly into healthy tissues, being evenly distributed within the body [166], so only a small amount of the administrated drug ends up reaching the target site, and its distribution into healthy tissues can lead to resistance. As seen above, gold nanoparticles can be used as vectors for targeting cancer tissue/cells so that the biodistribution of drugs is optimised. Several studies show their use as drug carriers [163, 183, 196-199, 202-211]. The administration of hydrophobic drugs require molecular encapsulation and nanoparticles are particularly efficient in evading the reticulo-endothelial system [210].

A recent cellular uptake study showed that the gellan gum-reduced gold nanoparticles were directed to cancer cells, while no uptake was observed in normal mouse embryonic fibroblast cells [202]. The toxicity was evaluated in rats showing that the animals did not have any hematological or biochemical abnormalities. A study on colloidal gold gastrointestinal uptake in mice revealed that smaller particles (4 nm) crossed the gastrointestinal tract more readily [212]. Some patents exist referring also the use of gold nanoparticles as fast detection sensors for drug residues [213-216].

Gene Silencing

Antisense DNA and RNA interference using small-interfering RNA are powerful useful tools to block gene function and for sequence-specific post transcriptional gene silencing, being very important to down regulation of specific gene expression in cancer cells [166]. Gold nanoparticles have shown potential as intracellular delivery vehicles for antisense oligonucleotides [217] and for therapeutic small-interfering RNA by providing protection against RNAses and ease of functionalization for selective targeting [218, 219].

Photothermal Agents

It is commonly accepted that, above 315 K, cell viability is strongly reduced [166]. Therefore, a temperature increase, called hyperthermia, can induce apoptosis and even cell death and tissue ablation, if it goes above 323 K [166, 220]. Hyperthermia therapy in cancer has been widely used either via direct irradiation or through the use suitable temperature vectors [166, 221]. Gold nanoparticles are being investigated as photothermal agents for in-vivo applications, showing to be effective for the inhibition of pathogenic bacteria cell growth, including Gram-positive bacteria, Gram-negative bacteria and antibiotic-resistant bacteria [222], and have demonstrated effective photothermal destruction of cancer cells and tissue [166, 223-225]. After intravenous injection, the Au nanoparticles accumulated in skin tumor cells after 4-5 h and induced apoptotic death of tumor cells, completely inhibiting the tumor growth after just five days of treatment [225]. The use of non-invasive radiowaves at 13.56 MHz have also been shown to induce heat in Au nanoparticles and thermally destroy tumor tissue [226].

Radiotherapy

Radiotherapy uses ionizing radiation to fight the proliferation of cancer cells [166]. However, sparing healthy tissues around a tumor is very complicated. Gold nanopartices can act as antennas, providing enhanced radiation targeting with lower radiation doses, consequently avoiding damage to healthy tissues. The use of this strategy has led to improvement in the treatment on cancer cells sparing the normal surrounding tissues in mice models [227] and in breast cancer [228].

Alzheimer's Disease

Alzheimer's disease is a neurodegenerative disease that affects near 5 million people and causes shrinkage of brain tissue (Figure 7), being the most common cause of dementia [229]. The beta-amyloid toxicity has been hypothesized to initiate the pathogenesis of Alzheimer's disease. The characteristic fibrillar morphology of amyloid-beta aggregates, that constitute the main components of senile plaque, has long been considered to account for the neurotoxicity. It has been seen that certain amyloid fibril morphologies are more toxic than others. Gold nanoparticles have been used to visualize by electron microscopy positive-charge clusters on beta-amyloid fibrils, showing that the surface structure of the fibrils plays an important role in amyloid toxicity [230]. It also has been shown that gold nanoparticles can destroy the beta-amyloid fibrils and plaque associated with Alzheimer's disease [229, 231-

234]. Other possibilities exist for numerous similar radiative applications of gold [235].

Figure 7. Cross-sections of a normal young brain (left) and senior brain (right), showing shrinkage of brain tissue caused by Alzheimer's disease.

Recently, it is known that metallic gold releases charged gold atoms when placed intra-cerebrally and gold ions can affect inflammation in the brain [236]. This suggests that metallic Au can be also used as a safe suppressor of inflammation in the central nervous system [236].

FUTURE PROSPECTS

As discussed above, research on gold nanoparticles shows its importance in several applications. The potential of gold for research is enormous and certainly future results will lead to more practical, commercial applications, not yet possible to foresee.

ACKNOWLEDGMENTS

I am grateful to to Mr. Jeffrey W. Hamilton for the permission to use the illustration of Moses breaking the Tablets of the Law, taken from the website http://lavistachurchofchrist.org/. My acknowledgement also to Fundação para a Ciência e a Tecnologia (FCT) for financial support (CIENCIA 2007 program and project PTDC/QUI-QUI/100682/2008, financed by FCT and FEDER in the context of program COMPETE).

REFERENCES

[1] Haruta, M. Size- and support-dependency in the catalysis of gold. *Catal. Today*. 1997, 36, 153-166.

[2] Bond, GC; Thompson, DT. Catalysis by gold. *Catal. Rev. Sci. Eng.* 1999, 41, 319-388.

[3] Haruta, M. Size- and support-dependency in the catalysis of noble metals: Gold as a typical example. *Physics and Chemistry of Clusters.* 2001, 117, 99-109.

[4] Haruta, M; Date, M. Advances in the catalysis of Au nanoparticles. *Appl. Catal. A: Gen.* 2001, 222, 427-437.

[5] Haruta, M. Catalysis of gold nanoparticles deposited on metal oxides. *Cattech.* 2002, 6, 102-115.

[6] Haruta, A. When gold is not noble: Catalysis by nanoparticles. *Chem. Rec.* 2003, 3, 75-87.

[7] Haruta, M. Gold as a novel catalyst in the 21st century: Preparation, working mechanism and applications. *Gold Bull.* 2004, 37, 27-36.

[8] Hashmi, ASK; Hutchings, GJ. Gold catalysis. *Angew. Chem. Int. Ed.* 2006, 45, 7896-7936.

[9] Bond, GC; Louis, C; Thompson, DT. Catalysis by Gold. Hutchings, GJ, editor. vol. 6. London, United Kingdom: *Imperial College Press;* 2006.

[10] Carabineiro, SAC; Thompson, DT. Catalytic Applications for Gold Nanotechnology. In: Heiz, EU, Landman, U, editors. *Nanocatalysis.* vol. Berlin, Heidelberg, New York: Springer-Verlag; 2007. p. 377-489.

[11] Carabineiro, SAC; Thompson, DT. Gold Catalysis. In: Corti, C, Holliday, R, editors. *Gold: Science and Applications.* vol. Boca Raton, London, New York: CRC Press, Taylor and Francis Group; 2010. p. 89-122.

[12] Turkevich, J; Garton, G; Stevenson, PC. The colour of colloidal gold. *Journal of Colloid Science.* 1954, 9, 26-35.

[13] Turkevich, J. Colloidal gold. Part II. *Gold Bull.* 1985, 18, 125-131.

[14] Sharma, V; Park, K; Srinivasarao, M. Colloidal dispersion of gold nanorods: Historical background, optical properties, seed-mediated synthesis, shape separation and self-assembly. *Materials Science and Engineering R.* 2009, 65, 1-38.

[15] Carabineiro, SAC. Colloidal Gold. Colloids: Classification, Properties and Applications. vol.: Nova Publishers, Nova Science Pub Inc., in press; 2012.

[16] Bond, GC; Thompson, DT. Catalysis by Gold. *Catal. Rev. Sci. Eng.* 1999, 41, 319.

[17] Turkevich, J. Colloidal gold. Part I. Gold Bull. 1985, 18, 86-91.

[18] Svedberg, T. The Formation of Colloids. vol. New York: D. van Nostrand Company, Inc.; 1921.

[19] Svedberg, T; Tiselius, A. Colloid Chemistry. vol. New York: The Chemical Catalog Company, Inc.; 1928.

[20] Saraiva, SM; de Oliveira, JF. Control of Particle Size in the Preparation of Colloidal Gold. *Journal of Dispersion Science and Technology.* 2002, 23, 837-844.

[21] Leff, DV; Ohara, PC; Heath, JR; Gelbart, WM. Thermodynamic Control of Gold Nanocrystal Size: Experiment and Theory. *The Journal of Physical Chemistry.* 1995, 99, 7036-7041.

[22] Turkevich, J; Stevenson, PC; Hillier, J. A study of the nucleation and growth processes in the synthesis of colloidal gold. *Discussions of the Faraday Society.* 1951, 11, 55-75.

[23] Kimling, J; Maier, M; Okenve, B; Kotaidis, V; Ballot, H; Plech, A. Turkevich Method for Gold Nanoparticle Synthesis Revisited. *The Journal of Physical Chemistry B.* 2006, 110, 15700-15707.

[24] Frens, G. Particle size and sol stability in metal colloids. *Colloid & Polymer Science.* 1972, 250, 736-741.

[25] Frens, G. Controlled Nucleation for the Regulation of the Particle Size in Monodisperse Gold Suspensions. *Nature Physical Science.* 1973, 241, 20-22.

[26] Brust, M; Walker, M; Bethell, D; Schiffrin, DJ; Whyman, R. Synthesis of Thiol-derivatised Gold Nanoparticles in a Two-phase Liquid-Liquid System. *Chemical Communications.* 1994, 801-802.

[27] Martin, MN; Basham, JI; Chando, P; Eah, S-K. Charged gold nanoparticles in non-polar solvents: 10-min synthesis and 2D self-assembly. *Langmuir.* 2010, 26, 7410-7417.

[28] Perrault, SD; Chan, WCW. Synthesis and Surface Modification of Highly Monodispersed, Spherical Gold Nanoparticles of 50-200 nm. *Journal of the Americal Chemical Society.* 2009, 131, 17042-17043.

[29] Park, J-E; Atobe, M; Fuchigami, T. Synthesis of multiple shapes of gold nanoparticles with controlled sizes in aqueous solution using ultrasound. *Ultrasonics Sonochemistry.* 2006, 13, 237-241.

[30] Yang, S; Zhang, T; Zhang, L; Wang, S; Yang, Z; Ding, B. Continuous synthesis of gold nanoparticles and nanoplates with controlled size and

shape under UV irradiation. *Colloids and Surfaces A: Physicochemical and Engineering Aspects.* 2007, 296, 37-44.
[31] Lee, S-J; Gavriilidis, A. Supported Au Catalysts for Low-Temperature CO Oxidation Prepared by Impregnation. *J. Catal.* 2002, 206, 305-313.
[32] Carabineiro, SAC; Silva, AMT; Dražić, G; Tavares, PB; Figueiredo, JL. Effect of chloride on the sinterization of Au/CeO_2 catalysts. *Catal. Today.* 2010, 154, 293-302.
[33] Meyer, R; Lemire, C; Shaikhutdinov, SK; Freund, H. Surface chemistry of catalysis by gold. *Gold Bull.* 2004, 37, 72-+.
[34] Carabineiro, SAC; Silva, AMT; Dražić, G; Tavares, PB; Figueiredo, JL. Gold nanoparticles on ceria supports for the oxidation of carbon monoxide. *Catal. Today.* 2010, 154, 21–30.
[35] Xu, Q; Kharas, KCC; Datye, AK. The preparation of highly dispersed Au/Al2O3 by aqueous impregnation. *Catal. Lett.* 2003, 85, 229-235.
[36] Bowker, M; Nuhu, A; Soares, J. High activity supported gold catalysts by incipient wetness impregnation. *Catal. Today.* 2007, 122, 245-247.
[37] Carabineiro, SAC; Bastos, SST; Orfao, JJM; Pereira, MFR; Delgado, JJ; Figueiredo, JL. Carbon Monoxide Oxidation Catalysed by Exotemplated Manganese Oxides. *Catal. Lett.* 2010, 134, 217-227.
[38] Carabineiro, SAC; Bastos, SST; Órfão, JJM; Pereira, MFR; Delgado, JJ; Figueiredo, JL. Exotemplated ceria catalysts with gold for CO oxidation. *Appl. Catal. A: Gen.* 2010, 381, 150-160.
[39] Carabineiro, SAC; Machado, BF; Bacsa, RR; Serp, P; Dražić, G; Faria, JL; Figueiredo, JL. Catalytic performance of Au/ZnO nanocatalysts for CO oxidation. *J. Catal.* 2010, 273, 191–198.
[40] Carabineiro, SAC; Silva, AMT; Drazic, G; Figueiredo, JL. Preparation of Au nanoparticles on Ce-Ti-O supports. In: E.M. Gaigneaux, M. Devillers, S. Hermans, P. A. Jacobs, J. A. Martens, Ruiz, P, editors. *Studies in Surface Science and Catalysis.* vol. Volume 175: Elsevier; 2010. p. 457-461.
[41] Carabineiro, SAC; Bogdanchikova, N; Avalos-Borja, M; Pestryakov, A; Tavares, P; Figueiredo, JL. Gold supported on metal oxides for carbon monoxide oxidation. *Nano Res.* 2011, 4, 180–193.
[42] Carabineiro, S; Bogdanchikova, N; Pestryakov, A; Tavares, P; Fernandes, L; Figueiredo, J. Gold nanoparticles supported on magnesium oxide for CO oxidation. *Nanoscale Research Letters.* 2011, 6, 435 (436 pages).
[43] Lee, S-J; Gavriilidis, A; Pankhurst, QA; Kyek, A; Wagner, FE; Wong, PCL; Yeung, KL. Effect of Drying Conditions of Au–Mn Co-

Precipitates for Low-Temperature CO Oxidation. *J. Catal.* 2001, 200, 298-308.

[44] Seker, E; Gulari, E. Single step sol–gel made gold on alumina catalyst for selective reduction of NOx under oxidizing conditions: effect of gold precursor and reaction conditions. *Appl. Catal. A: Gen.* 2002, 232, 203-217.

[45] Wolf, A; Schüth, F. A systematic study of the synthesis conditions for the preparation of highly active gold catalysts. *Appl. Catal. A: Gen.* 2002, 226, 1-13.

[46] Wang, DH; Hao, ZP; Cheng, DY; Shi, XC; Hu, C. Influence of pretreatment conditions on low-temperature CO oxidation over Au/MOx/Al2O3 catalysts. *J. Mol. Catal. A: Chem.* 2003, 200, 229-238.

[47] Kung, HH; Kung, MC; Costello, CK. Supported Au catalysts for low temperature CO oxidation. *J. Catal.* 2003, 216, 425-432.

[48] Moreau, F; Bond, GC; Taylor, AO. Gold on titania catalysts for the oxidation of carbon monoxide: control of pH during preparation with various gold contents. *J. Catal.* 2005, 231, 105-114.

[49] Zanella, R; Delannoy, L; Louis, C. Mechanism of deposition of gold precursors onto TiO2 during the preparation by cation adsorption and deposition-precipitation with NaOH and urea. *Appl. Catal. A: Gen.* 2005, 291, 62-72.

[50] Dobrosz, I; Jiratova, K; Pitchon, V; Rynkowski, JM. Effect of the preparation of supported gold particles on the catalytic activity in CO oxidation reaction. *J. Mol. Catal. A: Chem.* 2005, 234, 187-197.

[51] Gluhoi, AC; Nieuwenhuys, BE. Structural and chemical promoter effects of alkali (earth) and cerium oxides in CO oxidation on supported gold. *Catal. Today.* 2007, 122, 226-232.

[52] Hugon, A; El Kolli, N; Louis, C. Advances in the preparation of supported gold catalysts: Mechanism of deposition, simplification of the procedures and relevance of the elimination of chlorine. *J. Catal.* 2010, 274, 239-250.

[53] Prati, L; Martra, G. New gold catalysts for liquid phase oxidation. *Gold Bull.* 1999, 32, 96-101.

[54] Lin, J-N; Wan, B-Z. Effects of preparation conditions on gold/Y-type zeolite for CO oxidation. *Appl. Catal. B: Environ.* 2003, 41, 83-95.

[55] Kang, YM; Wan, BZ. Preparation of gold in Y-type zeolite for carbon-monoxide oxidation. *Appl. Catal. A: Gen.* 1995, 128, 53-60.

[56] Kang, YM; Wan, BZ. Gold and iron supported on Y-type zeolite for carbon monoxide oxidation. *Catal. Today.* 1997, 35, 379-392.

[57] Horvath, D; Polisset-Thfoin, M; Fraissard, J; Guczi, L. Novel preparation method and characterization of Au-Fe/HY zeolite containing highly stable gold nanoparticles inside zeolite supercages. *Solid State Ionics.* 2001, 141, 153-156.

[58] Chen, JH; Lin, JN; Kang, YM; Yu, WY; Kuo, CN; Wan, BZ. Preparation of nano-gold in zeolites for CO oxidation: Effects of structures and number of ion exchange sites of zeolites. *Appl. Catal. A: Gen.* 2005, 291, 162-169.

[59] Magadzu, T; Hearne, GR; Scurrell, MS. Characteristics of gold-zeolite Y catalysts in CO oxidation and ethylene hydrogenation. In: Xu, R, Gao, Z, Chen, J, Yan, W, editors. From Zeolites to Porous Mof Materials: The 40th Anniversary of International Zeolite Conference, *Proceedings of the 15th International Zeolite Conference.* vol. 1702007. p. 1059-1064.

[60] Bogdanchikova, N; Simakov, A; Smolentseva, E; Pestryakov, A; Farias, MH; Diaz, JA; Tompos, A; Avalos, M. Stabilization of catalytically active gold species in Fe-modified zeolites. *Applied Surface Science.* 2008, 254, 4075-4083.

[61] Ivanova, S; Petit, C; Pitchon, V. A new preparation method for the formation of gold nanoparticles on an oxide support. *Appl. Catal. A: Gen.* 2004, 267, 191-201.

[62] Ivanova, S; Petit, C; Pitchon, V. Application of heterogeneous gold catalysis with increased durability: Oxidation of CO and hydrocarbons at low temperature. *Gold Bull.* 2006, 39, 3-8.

[63] Ivanova, S; Pitchon, V; Petit, C. Application of the direct exchange method in the preparation of gold catalysts supported on different oxide materials. *J. Mol. Catal. A: Chem.* 2006, 256, 278-283.

[64] Dobrosz-Gomez, I; Kocemba, I; Rynkowski, JM. Factors influencing structure and catalytic activity of Au/Ce1-xZrxO2 catalysts in CO oxidation. *Appl. Catal. B: Environ.* 2009, 88, 83-97.

[65] Azizi, Y; Pitchon, V; Petit, C. Effect of support parameters on activity of gold catalysts: Studies of ZrO2, TiO2 and mixture. *Appl. Catal. A: Gen.* 2010, 385, 170-177.

[66] Fisher, JM. Beware! Fulminating gold. *Chemistry in Britain.* 2003, 39, 12-12.

[67] Steinhauser, G; Evers, J; Jakob, S; Klapotke, TM; Oehlinger, G. A review on fulminating gold (Knallgold). *Gold Bull.* 2008, 41, 305-317.

[68] Sunagawa, Y; Yamamoto, K; Takahashi, H; Muramatsu, A. Liquid-phase reductive deposition as a novel nanoparticle synthesis method and

its application to supported noble metal catalyst preparation. *Catal. Today.* 2008, 132, 81-87.

[69] L. Brey; T. E. Wood; G.M. Buccellato; M.E. Jones; C.S. Chamberlain; A.R. Siedle, Catalysts, activating agents, support media, and related methodologies useful for making catalyst systems especially when the catalyst is deposited onto the support media using Physical Vapor Deposition, 3M Innovative Properties Co, WO Patent 2005/030382.

[70] J.T. Brady; M.E. Jones; L.A. Brey; G.M. Buccellato; C.S. Chamberlain; J.S. Huberty; A.R. Siedle; T.E. Wood, Heterogeneous, composite carbonaceous catalyst system and methods that use catalytically active gold, 3M Innovative Properties Co, WO Patent 2006/074126.

[71] Yuan, YZ; Kozlova, AP; Asakura, K; Wan, HL; Tsai, K; Iwasawa, Y. Supported Au catalysts prepared from Au phosphine complexes and As-precipitated metal hydroxides: Characterization and low-temperature CO oxidation. *J. Catal.* 1997, 170, 191-199.

[72] Kozlova, AP; Sugiyama, S; Kozlov, AI; Asakura, K; Iwasawa, Y. Iron-oxide supported gold catalysts derived from gold-phosphine complex Au(PPh3)(NO3): State and structure of the support. *J. Catal.* 1998, 176, 426-438.

[73] Kozlova, AP; Kozlov, AI; Sugiyama, S; Matsui, Y; Asakura, K; Iwasawa, Y. Study of gold species in iron-oxide-supported gold catalysts derived from gold-phosphine complex Au(PPh3)(NO3) and As-precipitated wet Fe(OH)(3). *J. Catal.* 1999, 181, 37-48.

[74] Kozlov, AI; Kozlova, AP; Asakura, K; Matsui, Y; Kogure, T; Shido, T; Iwasawa, Y. Supported gold catalysts prepared from a gold phosphine precursor and As-precipitated metal-hydroxide precursors: Effect of preparation conditions on the catalytic performance. *J. Catal.* 2000, 196, 56-65.

[75] Okumura, M; Tsubota, S; Haruta, M. Preparation of supported gold catalysts by gas-phase grafting of gold acethylacetonate for low-temperature oxidation of CO and of H2. *Journal of Molecular Catalysis A: Chemical.* 2003, 199, 73-84.

[76] Bond, GC; Thompson, DT. Gold-catalysed oxidation of carbon monoxide. *Gold Bull.* 2000, 33, 41-51.

[77] Bond, G; Thompson, D. Formulation of mechanisms for gold-catalysed reactions. *Gold Bull.* 2009, 42, 247-259.

[78] Date, M; Haruta, M. Moisture effect on CO oxidation over Au/TiO2 catalyst. *J. Catal.* 2001, 201, 221-224.

[79] Date, M; Okumura, M; Tsubota, S; Haruta, M. Vital role of moisture in the catalytic activity of supported gold nanoparticles. *Angew. Chem. Int.* Ed. 2004, 43, 2129-2132.

[80] Cameron, D; Holliday, R; Thompson, D. Gold's future role in fuel cell systems. *Journal of Power Sources*. 2003, 118, 298-303.

[81] Kandoi, S; Gokhale, AA; Grabow, LC; Dumesic, JA; Mavrikakis, M. Why Au and Cu are more selective than Pt for preferential oxidation of CO at low temperature. *Catal. Lett.* 2004, 93, 93-100.

[82] Panzera, G; Modafferi, V; Candamano, S; Donato, A; Frusteri, F; Antonucci, PL. CO selective oxidation on ceria-supported Au catalysts for fuel cell application. *Journal of Power Sources*. 2004, 135, 177-183.

[83] Tabakova, T; Idakiev, V; Tenchev, K; Boccuzzi, F; Manzoli, M; Chiorino, A. Pure hydrogen production on a new gold-thoria catalyst for fuel cell applications. *Appl. Catal. B: Environ.* 2006, 63, 94-103.

[84] Tada, M; Bal, R; Mu, X; Coquet, R; Namba, S; Iwasawa, Y. Low-temperature PROX (preferential oxidation) on novel CeO2-supported Cu-cluster catalysts under fuel-cell operating conditions. *Chemical Communications*. 2007, 4689-4691.

[85] Imai, H; Date, M; Tsubota, S. Preferential oxidation of CO in H(2)-Rich gas at low temperatures over Au nanoparticles supported on metal oxides. *Catal. Lett.* 2008, 124, 68-73.

[86] Kim, YH; Park, ED; Lee, HC; Lee, D. Selective CO removal in a H-2-rich stream over supported Ru catalysts for the polymer electrolyte membrane fuel cell (PEMFC). *Appl. Catal. A: Gen.* 2009, 366, 363-369.

[87] Park, ED; Lee, D; Lee, HC. Recent progress in selective CO removal in a H-2-rich stream. *Catal. Today*. 2009, 139, 280-290.

[88] Pitois, A; Pilenga, A; Pfrang, A; Tsotridis, G. Temperature-dependent CO desorption kinetics on supported gold nanoparticles: Relevance to clean hydrogen production and fuel cell systems. *International Journal of Hydrogen Energy*. 2011, 36, 4375-4385.

[89] Andreeva, D; Idakiev, V; Tabakova, T; Andreev, A. Low-temperature water-gas shift reaction over Au/alpha-Fe2O3. *J. Catal.* 1996, 158, 354-355.

[90] Tabakova, T; Idakiev, V; Andreeva, D; Mitov, I. Influence of the microscopic properties of the support on the catalytic activity of Au/ZnO, Au/ZrO2, Au/Fe2O3, Au/Fe2O3-ZnO, Au/Fe2O3-ZrO2 catalysts for the WGS reaction. *Appl. Catal. A: Gen.* 2000, 202, 91-97.

[91] Andreeva, D. Low temperature water gas shift over gold catalysts. *Gold Bull.* 2002, 35, 82-88.

[92] Andreeva, D; Idakiev, V; Tabakova, T; Ilieva, L; Falaras, P; Bourlinos, A; Travlos, A. Low-temperature water-gas shift reaction over Au/CeO2 catalysts. *Catal. Today.* 2002, 72, 51-57.
[93] Andreeva, D; Ivanova, I; Ilieva, L; Abrashev, MV. Gold catalysts supported on ceria and ceria-alumina for water-gas shift reaction. *Appl. Catal. A: Gen.* 2006, 302, 127-132.
[94] Andreeva, D; Ivanov, I; Ilieva, L; Sobczak, JW; Avdeev, G; Tabakova, T. Nanosized gold catalysts supported on ceria and ceria-alumina for WGS reaction: Influence of the preparation method. *Appl. Catal. A: Gen.* 2007, 333, 153-160.
[95] Andreeva, D; Ivanov, I; Ilieva, L; Abrashev, MV; Zanella, R; Sobczak, JW; Lisowski, W; Kantcheva, M; Avdeev, G; Petrov, K. Gold catalysts supported on ceria doped by rare earth metals for water gas shift reaction: Influence of the preparation method. *Appl. Catal. A: Gen.* 2009, 357, 159-169.
[96] Prati, L; Rossi, M. Gold on carbon as a new catalyst for selective liquid phase oxidation of diols. *J. Catal.* 1998, 176, 552-560.
[97] Bianchi, C; Porta, F; Prati, L; Rossi, M. Selective liquid phase oxidation using gold catalysts. *Top. Catal.* 2000, 13, 231-236.
[98] Biella, S; Castiglioni, GL; Fumagalli, C; Prati, L; Rossi, M. Application of gold catalysts to selective liquid phase oxidation. *Catal. Today.* 2002, 72, 43-49.
[99] Biella, S; Prati, L; Rossi, M. Selectivity control in the oxidation of phenylethane-1,2-diol with gold catalyst. *Inorganica Chimica Acta.* 2003, 349, 253-257.
[100] Biella, S; Prati, L; Rossi, M. Gold catalyzed oxidation of aldehydes in liquid phase. *J. Mol. Catal. A: Chem.* 2003, 197, 207-212.
[101] Biella, S; Rossi, M. Gas phase oxidation of alcohols to aldehydes or ketones catalysed by supported gold. *Chemical Communications.* 2003, 378-379.
[102] Porta, F; Rossi, M. Gold nanostructured materials for the selective liquid phase catalytic oxidation. *J. Mol. Catal. A: Chem.* 2003, 204, 553-559.
[103] Comotti, M; Della Pina, C; Matarrese, R; Rossi, M. The catalytic activity of "Naked" gold particles. *Angew. Chem. Int. Ed.* 2004, 43, 5812-5815.
[104] Comotti, M; Della Pina, C; Falletta, E; Rossi, M. Is the biochemical route always advantageous? The case of glucose oxidation. *J. Catal.* 2006, 244, 122-125.

[105] Comotti, M; Della Pina, C; Matarrese, R; Rossi, M; Siani, A. Oxidation of alcohols and sugars using Au/C catalysts. Part 2. Sugars (vol 291, pg 204, 2005). *Appl. Catal. A: Gen.* 2006, 301, 143-143.

[106] Della Pina, C; Falletta, E; Rossi, M. Selective oxidation of tertiary amines on gold catalysts. *Top. Catal.* 2007, 44, 325-329.

[107] Pagliaro, M; Ciriminna, R; Kimura, H; Rossi, M; Della Pina, C. From glycerol to value-added products. *Angew. Chem. Int. Ed.* 2007, 46, 4434-4440.

[108] Della Pina, C; Falletta, E; Prati, L; Rossi, M. Selective oxidation using gold. *Chem. Soc. Rev.* 2008, 37, 2077-2095.

[109] Pagliaro, M; Ciriminna, R; Kimura, H; Rossi, M; Della Pina, C. Recent advances in the conversion of bioglycerol into value-added products. *European Journal of Lipid Science and Technology.* 2009, 111, 788-799.

[110] Rodrigues, EG; Carabineiro, SAC; Delgado, JJ; Chen, X; Pereira, MFR; Orfao, JJM. Gold supported on carbon nanotubes for the selective oxidation of glycerol. *J. Catal.* 2012, 285, 83-91.

[111] Della Pina, C; Falletta, E; Rossi, M. Update on selective oxidation using gold. *Chem. Soc. Rev.* 2012, 41, 350-369.

[112] Hutchings, GJ. Gold catalysis in chemical processing. *Catal. Today.* 2002, 72, 11-17.

[113] Milone, C; Tropeano, ML; Gulino, G; Neri, G; Ingoglia, R; Galvagno, S. Selective liquid phase hydrogenation of citral on Au/Fe2O3 catalysts. *Chemical Communications.* 2002, 868-869.

[114] Milone, C; Ingoglia, R; Pistone, A; Neri, G; Frusteri, F; Galvagno, S. Selective hydrogen of alpha, beta-unsaturated ketones to alpha, beta-unsaturated alcohols on gold-supported catalysts. *J. Catal.* 2004, 222, 348-356.

[115] Milone, C; Ingoglia, R; Schipilliti, L; Crisafulli, C; Neri, G; Galvagno, S. Selective hydrogenation of alpha,beta-unsaturated ketone to alpha,beta-unsaturated alcohol on gold-supported iron oxide catalysts: Role of the support. *J. Catal.* 2005, 236, 80-90.

[116] Claus, P. Heterogeneously catalysed hydrogenation using gold catalysts. *Appl. Catal. A: Gen.* 2005, 291, 222-229.

[117] Milone, C; Ingoglia, R; Galvagno, S. Gold supported on iron oxy-hydroxides: a versatile tool for the synthesis of fine chemicals. *Gold Bull.* 2006, 39, 54-65.

[118] Milone, C; Crisafulli, C; Ingoglia, R; Schipilliti, L; Galvagno, S. A comparative study on the selective hydrogenation of alpha,beta

unsaturated aldehyde and ketone to unsaturated alcohols on Au supported catalysts. *Catal. Today.* 2007, 122, 341-351.

[119] Kim, S; Bae, SW; Lee, JS; Park, J. Recyclable gold nanoparticle catalyst for the aerobic alcohol oxidation and C-C bond forming reaction between primary alcohols and ketones under ambient conditions. *Tetrahedron.* 2009, 65, 1461-1466.

[120] Lenz, J; Campo, BC; Alvarez, M; Volpe, MA. Liquid phase hydrogenation of alpha,beta-unsaturated aldehydes over gold supported on iron oxides. *J. Catal.* 2009, 267, 50-56.

[121] Wang, C-M; Fan, K-N; Liu, Z-P. Oxide-supported single gold catalyst for selective hydrogenation of acrolein predicted from first principles. *J. Catal.* 2009, 266, 343-350.

[122] Yang, Q-Y; Zhu, Y; Tian, L; Pei, Y; Qiao, M-H; Fan, K-N. Influence of Au/TiO2 Catalyst Preparation Parameters on the Selective Hydrogenation of Crotonaldehyde. *Acta Physico-Chimica Sinica.* 2009, 25, 1853-1860.

[123] You, K-J; Chang, C-T; Liaw, B-J; Huang, C-T; Chen, Y-Z. Selective hydrogenation of alpha,beta-unsaturated aldehydes over Au/MgxAlO hydrotalcite catalysts. *Appl. Catal. A: Gen.* 2009, 361, 65-71.

[124] Gluhoi, AC; Bakker, JW; Nieuwenhuys, BE. Gold, still a surprising catalyst: Selective hydrogenation of acetylene to ethylene over Au nanoparticles. *Catal. Today.* 2010, 154, 13-20.

[125] McEwan, L; Julius, M; Roberts, S; Fletcher, JCQ. A review of the use of gold catalysts in selective hydrogenation reactions. *Gold Bull.* 2010, 43, 298-306.

[126] Ide, MS; Hao, B; Neurock, M; Davis, RJ. Mechanistic Insights on the Hydrogenation of alpha,beta-Unsaturated Ketones and Aldehydes to Unsaturated Alcohols over Metal Catalysts. *ACS Catal.* 2012, 2, 671-683.

[127] Ilieva, L; Pantaleo, G; Ivanov, I; Venezia, AM; Andreeva, D. Gold catalysts supported on CeO2 and CeO2-Al2O3 for NOx reduction by CO. *Appl. Catal. B: Environ.* 2006, 65, 101-109.

[128] Nga, NLT; Potvin, C; Djega-Mariadassou, G; Delannoy, L; Louis, C. Catalytic reduction of nitrogen monoxide by propene in the presence of excess oxygen over gold based ceria catalyst. *Top. Catal.* 2007, 42-43, 91-94.

[129] Ilieva-Gencheva, L; Pantaleo, G; Mintcheva, N; Ivanov, I; Venezia, AM; Andreeva, D. Nano-structured gold catalysts supported on CeO2 and

CeO2-Al2O3 for NOx reduction by CO: Effect of catalyst pretreatment and feed composition. *J. Nanosci. Nanotechnol.* 2008, 8, 867-873.

[130] Nguyen, LQ; Salim, C; Hinode, H. Performance of nano-sized Au/TiO2 for selective catalytic reduction of NOx by propene. *Appl. Catal. A: Gen.* 2008, 347, 94-99.

[131] Miguel, P; Granger, P; Jagtap, N; Umbarkar, S; Dongare, M; Dujardin, C. NO reduction under diesel exhaust conditions over Au/Al2O3 prepared by deposition-precipitation method. *J. Mol. Catal. A: Chem.* 2010, 322, 90-97.

[132] de Almeida, MP; Carabineiro, SAC. Dioxin decomposition and detection using gold based materials. *Recent Patents on Chemical Engineering.* 2012, 5, 56-62.

[133] Minico, S; Scire, S; Crisafulli, C; Maggiore, R; Galvagno, S. Catalytic combustion of volatile organic compounds on gold/iron oxide catalysts. *Appl. Catal. B: Environ.* 2000, 28, 245-251.

[134] Centeno, MA; Paulis, M; Montes, M; Odriozola, JA. Catalytic combustion of volatile organic compounds on Au/CeO2/Al2O3 and Au/Al2O3 catalysts. *Appl. Catal. A: Gen.* 2002, 234, 65-78.

[135] Scire, S; Minico, S; Crisafulli, C; Satriano, C; Pistone, A. Catalytic combustion of volatile organic compounds on gold/cerium oxide catalysts. *Appl. Catal. B: Environ.* 2003, 40, 43-49.

[136] Roark, SE; Cabrera-Fonseca, J; Milazzo, MC; White, JH; Wander, JD. Catalytic oxidation of volatile organic liquids. *Journal of Environmental Engineering-Asce.* 2004, 130, 329-337.

[137] Centeno, MA; Paulis, M; Montes, M; Odriozola, JA. Catalytic combustion of volatile organic compounds on gold/titanium oxynitride catalysts. *Appl. Catal. B: Environ.* 2005, 61, 177-183.

[138] Della Pina, C; Dimitratos, N; Falletta, E; Rossi, M; Siani, A. Catalytic performance of gold catalysts in the total oxidation of VOCs. *Gold Bull.* 2007, 40, 67-72.

[139] Gennequin, C; Lamallem, M; Cousin, R; Siffert, S; Aissi, F; Aboukais, A. Catalytic oxidation of VOCs on Au/Ce-Ti-O. *Catal. Today.* 2007, 122, 301-306.

[140] Lamallem, M; El Ayadi, H; Gennequin, C; Cousin, R; Siffert, S; Aissi, F; Aboukais, A. Effect of the preparation method on Au/Ce-Ti-O catalysts activity for VOCs oxidation. *Catal. Today.* 2008, 137, 367-372.

[141] Li, C; Shen, Y; Jia, M; Sheng, S; Adebajo, MO; Zhu, H. Catalytic combustion of formaldehyde on gold/iron-oxide catalysts. *Catal. Commun.* 2008, 9, 355-361.

[142] Li, J; Li, W. Effect of preparation method on the catalytic activity of Au/CeO2 for VOCs oxidation. *Journal of Rare Earths.* 2010, 28, 547-551.

[143] Santos, VP; Carabineiro, SAC; Tavares, PB; Pereira, MFR; Orfao, JJM; Figueiredo, JL. Oxidation of CO, ethanol and toluene over TiO2 supported noble metal catalysts. *Appl. Catal. B: Environ.* 2010, 99, 198-205.

[144] Bastos, SST; Carabineiro, SAC; Orfao, JJM; Pereira, MFR; Delgado, JJ; Figueiredo, JL. Total oxidation of ethyl acetate, ethanol and toluene catalyzed by exotemplated manganese and cerium oxides loaded with gold. *Catal. Today.* 2012, 180, 148-154.

[145] Solsona, B; Perez-Cabero, M; Vazquez, I; Dejoz, A; Garcia, T; Alvarez-Rodriguez, J; El-Haskouri, J; Beltran, D; Amoros, P. Total oxidation of VOCs on Au nanoparticles anchored on Co doped mesoporous UVM-7 silica. *Chemical Engineering Journal.* 2012, 187, 391-400.

[146] Higby, GJ. Gold in Medicine - A review of its use in the West before 1900. *Gold Bull.* 1982, 15, 130-140.

[147] Kauffman, GB. The role of gold In alchemy. Part I. *Gold Bull.* 1985, 18, 31-44.

[148] Ray, PC. History of Hindu Chemistry. vol. Calcutta: The Bengal Chemical and Pharmaceutical Works Limited; 1903.

[149] Kauffman, GB. The role of gold In alchemy. *Part II. Gold Bull.* 1985, 18, 69-78.

[150] Chrestien, JA. De la Methode iatraleptique, ou observations pratiques sur l'efficacite des Remedes administres par la voie de l'Absorption cutanee dans le traitement de plusieurs Maladies internes et externes; et sur un Nouveau Remede dans le Traitement des Maladies veneriennes et lymphatiques vol. Paris: Croullebois & Crochard; 1811.

[151] Cheesman, J. On the Medicinal Properties of Gold, New York: Queen's College; 1812.

[152] Strug, DL; Priyadarsini, S; Hyman, MM, editors. Alcohol interventions: historical and sociocultural approaches. New York: The Haworth Press Inc.; 1986.

[153] Ochsner, EH. Colloidal Gold in Inoperable Cancer. *Clinical Medicine & Surgery.* 1935, 42, 321-324.

[154] Quimby, EH; Feitelberg, S; Silver, S. Radioactive Isotopes in Medicine and Biology: Medicine. vol. Philadelphia, PA: Lea & Febiger; 1962.

[155] Ansell, BM; Crook, A; Bywaters, EG; Mallard, JR. Evaluation of intra-articular colloidal gold au-198 in treatment of persistent knee effusions. *Annals of the Rheumatic Diseases*. 1963, 22, 435-&.
[156] Grahame, R; Ramsey, NW; Scott, JT. Radioactive colloidal gold - in chronic knee effusions with bakers cyst formation. *Annals of the Rheumatic Diseases*. 1970, 29, 159-&.
[157] Cayla, J. Colloidal Au-198 in the treatment of hydrarthrosis in rheumatoid polyarthritis. *Revue du rhumatisme et des maladies osteo-articulaires*. 1967, 34, 438-441.
[158] Correns, HJ; Zielke, A; Unverricht, A; Stiller, KJ. Intra-articular treatment of chronic polyarthritis with coarse colloidal Au198. *Radiobiologia, radiotherapia*. 1969, 10, 505-509.
[159] Gheorghe, N; Stoia, I. On the therapeuic value of colloidal radioactive gold (Au198) in the treatment of certain rheumatismal diseases. *Revue roumaine de physiologie*. 1970, 7, 323-328.
[160] Riccabon.G; Junger, H. Treatment of rheumatoid synovitis by intra-articular administration of colloidal Au-198. *Nuclear-Medizin*. 1970, 9, 279-&.
[161] Zielke, A; Correns, HJ; Unverricht, A; Stiller, KJ. Intra-articular treatment of progressive chronic polyarthritis with colloidal radioactive gold (Au 198). *Das Deutsche Gesundheitswesen*. 1970, 25, 1013-1020.
[162] Visuthikosol, V; Kumpolpunth, S. Intra-articular radioactive colloidal gold (Au-198) in the treatment of rheumatoid-arthritis. *Journal of the Medical Association of Thailand*. 1981, 64, 419-427.
[163] Sonvico, F; Dubernet, C; Colombo, P; Couvreur, P. Metallic colloid nanotechnology, applications in diagnosis and therapeutics. *Curr. Pharm. Design*. 2005, 11, 2091-2105.
[164] Bawarski, WE; Chidlowsky, E; Bharali, DJ; Mousa, SA. Emerging nanopharmaceuticals. *Nanomed.-Nanotechnol. Biol. Med*. 2008, 4, 273-282.
[165] Choi, B; Yang, J; Jang, E; Suh, J; Huh, Y; Lee, K; Haam, S. Gold Nanostructures as Photothermal Therapy Agent for Cancer. *Anti-Cancer Agents in Medicinal Chemistry*. 2011, in press.
[166] Conde, J; Doria, G; Baptista, P. Review: Noble Metal Nanoparticles Applications in Cancer. *Journal of Drug Delivery*. 2012, 2012, Article ID 751075, 751012 pages.
[167] Swanson, JN. The value of repeated colloidal tests in rheumatoid arthritis. *Annals of the Rheumatic Diseases*. 1949, 8, 232-236.

[168] De Preux, T. Colloidal gold sulfide in therapy of chronic inflammatory rheumatism. *Praxis.* 1951, 40, 696-700.

[169] De Preux, T. Recent discoveries in rheumatoid arthritis therapy: colloidal gold sulfide. *Medecine et hygiene.* 1951, 9, 57.

[170] De Preux, T. Colloidal gold in therapy of inflammatory rheumatism. *Medecine et hygiene.* 1952, 10, 265-266.

[171] Ferond, M. Colloidal gold therapy in polyarthritis. *Revue du rhumatisme et des maladies osteo-articulaires.* 1953, 20, 652.

[172] Forestier, J; Thevenoz, F. Le sulture dor colloidal dans la therapeutique de la polyarthrite chronique evolutive. *Presse Medicale.* 1954, 62, 1056-1057.

[173] Colamussi, V; Longhini, C; Vallini, A; Antonioli, G; Bazzanini, G. Colloidal gold sulfate in the therapy of chronic evolutive polyarthritis. *Minerva medica.* 1967, 58, 3137-3158.

[174] Frizziero, L; Facchini, A; Zizzi, F. Personal experience with the effect of a treatment with colloidal gold sulfide in rheumatoid arthritis. *Reumatismo.* 1977, 29, 148-153.

[175] Brown, C; Bushell, G; Whitehouse, M; Agrawal, D; Tupe, S; Paknikar, K; Tiekink, E. Nanogoldpharmaceutics (i) The use of colloidal gold to treat experimentally-induced arthritis in rat models; (ii) Characterization of the gold in Swarna bhasma, a microparticulate used in traditional Indian medicine. *Gold Bull.* 2007, 40, 245-250.

[176] CY, T; AL, S; SY, C; YH, C; PC, C; MY, C; DH, C; CH, C; CR, W; CL, W. Amelioration of collagen-induced arthritis in rats by nanogold. *Arthritis & Rheumatism.* 2007, 56, 544-554.

[177] Metselaar, JM; van den Berg, WB; Holthuysen, AEM; Wauben, MHM; Storm, G; van Lent, P. Liposomal targeting of glucocorticoids to synovial lining cells strongly increases therapeutic benefit in collagen type II arthritis. *Annals of the Rheumatic Diseases.* 2004, 63, 348-353.

[178] Jæger, GT; Larsen, S; Søli, N; Moe, L. Two years follow-up study of the pain-relieving effect of gold bead implantation in dogs with hip-joint arthritis. *Acta Veterinaria Scandinavica.* 2007, 49, 9.

[179] Horisberger, M; Rosset, J. Colloidal gold, a useful marker for transmission and scanning electron microscopy. *Journal of Histochemistry & Cytochemistry.* 1977, 25, 295-305.

[180] Hunter, EE. Practical Electron Microscopy: A Beginner's Illustrated Guide, 2nd edition. vol. Cabridge, New York, Melbourne, Madrid, Cape Town, Singapore, São Paulo, Delhi: Cambridge University Press; 1993.

[181] Bozzola, JJ; Russell, LD. Electron Microscopy Principles and Techniques for Biologists, 2nd Edition. Publishers, JaB, editor. vol. Boston, Toronto, London, Singapore1998.

[182] Cai, QY; Kim, SH; Choi, KS; Kim, SY; Byun, SJ; Kim, KW; Park, SH; Juhng, SK; Yoon, KH. Colloidal gold nanoparticles as a blood-pool contrast agent for x-ray computed tomography in mice. *Invest. Radiol.* 2007, 42, 797-806.

[183] Aryal, S; Pilla, S; Gong, SQ. Multifunctional Nano-Micelles Formed by Amphiphilic Gold-Polycaprolactone-Methoxy Poly(ethylene glycol) (Au-PCL-MPEG) Nanoparticles for Potential Drug Delivery Applications. *J. Nanosci. Nanotechnol.* 2009, 9, 5701-5708.

[184] Manohar, S; Rayavarapu, R; Petersen, W; van Leeuwen, TG. Cell viability studies of PEG-thiol treated gold nanorods as optoacoustic contrast agents. In: Oraevsky, AA, Wang, LV, editors. *Photons Plus Ultrasound: Imaging and Sensing* 2009. vol. 7177. Bellingham: Spie-Int Soc Optical Engineering; 2009.

[185] Kojima, H; Mukai, Y; Yoshikawa, M; Kamei, K; Yoshikawa, T; Morita, M; Inubushi, T; Yamamoto, TA; Yoshioka, Y; Okada, N; Seino, S; Nakagawa, S. Simple PEG Conjugation of SPIO via an Au-S Bond Improves Its Tumor Targeting Potency as a Novel MR Tumor Imaging *Agent. Bioconjugate Chem.* 2010, 21, 1026-1031.

[186] Zhang, ZY; Ross, RD; Roeder, RK. Preparation of functionalized gold nanoparticles as a targeted X-ray contrast agent for damaged bone tissue. *Nanoscale.* 2010, 2, 582-586.

[187] Peng, C; Wang, H; Guo, R; Shen, MW; Cao, XY; Zhang, GX; Shi, XY; Donghua Univ, CT. Acetylation of Dendrimer-Entrapped Gold Nanoparticles: Synthesis, Stability, and X-Ray Attenuation Property. vol. Shanghai: Donghua Univ Press; 2010.

[188] Peng, C; Wang, H; Guo, R; Shen, M; Cao, X; Zhu, M; Zhang, G; Shi, X. Acetylation of Dendrimer-Entrapped Gold Nanoparticles: Synthesis, Stability, and X-ray Attenuation Properties. *Journal of Applied Polymer Science.* 2011, 119, 1673-1682.

[189] Romano, EL; Romano, M. Staphylococcal protein a bound to colloidal gold: A useful reagent to label antigen-antibody sites in electron microscopy. *Immunochemistry.* 14, 711-715.

[190] Horovitz, O; Tomoaia, G; Mocanu, A; Yupsanis, T; Tomoaia-Cotisel, M. Protein binding to gold colloids. *Gold Bull.* 2007, 40, 213-218.

[191] Geuze, HJ; Slot, JW; van der Ley, PA; Scheffer, RCT; Griffith, JM. Use of Colloidal Gold Particles in Double-Labeling Immunoelectron

Microscopy of Ultrathin Frozen Tissue Sections. *The Journal of Cell Biology* 1981, 89 653-665.

[192] Fetni, R; Drouin, R; Lemieux, N; Messier, P-E; Richer, C-L. Simultaneous visualization of chromosome bands and hybridization signal using colloidal-gold labeling in electron microscopy, . Dec 1;():. *Proceedings of the National Academy of Sciences of the United States of America.* 1991, 88, 10916-10920.

[193] Paciotti, GF; Kingston, DGI; Tamarkin, L. Colloidal Gold Nanoparticles: A Novel Nanoparticle Platform for Developing Multifunctional Tumor-Targeted Drug Delivery Vectors. *Drug Development Research.* 2006, 67, 47-54.

[194] Visaria, R; Bischof, JC; Loren, M; Williams, B; Ebbini, E; Paciotti, G; Griffin, R. Nanotherapeutics for enhancing thermal therapy of cancer. *Int. J. Hyperthermia.* 2007, 23, 501-511.

[195] Qian, X; Peng, X-H; Ansari, DO; Yin-Goen, Q; Chen, GZ; Shin, DM; Yang, L; Young, AN; Wang, MD; Nie, S. In vivo tumor targeting and spectroscopic detection with surface-enhanced Raman nanoparticle tags. *Nat Biotech.* 2008, 26, 83-90.

[196] Subramani, K; Hosseinkhani, H; Khraisat, A; Hosseinkhani, M; Pathak, Y. Targeting Nanoparticles as Drug Delivery Systems for Cancer Treatment. *Curr. Nanosci.* 2009, 5, 135-140.

[197] Zhang, GD; Yang, Z; Lu, W; Zhang, R; Huang, Q; Tian, M; Li, L; Liang, D; Li, C. Influence of anchoring ligands and particle size on the colloidal stability and in vivo biodistribution of polyethylene glycol-coated gold nanoparticles in tumor-xenografted mice. *Biomaterials.* 2009, 30, 1928-1936.

[198] Park, C; Youn, H; Kim, H; Noh, T; Kook, YH; Oh, ET; Park, HJ; Kim, C. Cyclodextrin-covered gold nanoparticles for targeted delivery of an anti-cancer drug. *J. Mater. Chem.* 2009, 19, 2310-2315.

[199] Arvizo, R; Bhattacharya, R; Mukherjee, P. Gold nanoparticles: opportunities and challenges in nanomedicine. *Expert Opin. Drug Deliv.* 2010, 7, 753-763.

[200] Tam, NCM; Scott, BMT; Wilson, BC; Zheng, G. Lung cancer targeted Raman active phospholipid gold nanoparticles for ultrasensitive and specific molecular imaging and detection. In: Sterenborg, HVIA, editor. *Novel Biophotonic Techniques and Applications.* vol. 80902011.

[201] Wang, C-H; Liu, C-J; Chien, C-C; Chen, H-T; Hua, T-E; Leng, W-H; Chen, H-H; Kempson, IM; Hwu, Y; Hsiao, M; Lai, T-C; Wang, JL; Yang, C-S; Lin, H-M; Chen, Y-J; Margaritondo, G. X-ray synthesized

PEGylated (polyethylene glycol coated) gold nanoparticles in mice strongly accumulate in tumors. *Materials Chemistry and Physics.* 2011, 126, 352-356.

[202] Dhar, S; Mali, V; Bodhankar, S; Shiras, A; Prasad, BLV; Pokharkar, V. Biocompatible gellan gum-reduced gold nanoparticles: cellular uptake and subacute oral toxicity studies. *J. Appl. Toxicol.* 2011, 31, 411-420.

[203] Bhumkar, DR; Joshi, HM; Sastry, M; Pokharkar, VB. Chitosan reduced gold nanoparticles as novel carriers for transmucosal delivery of insulin. *Pharm. Res.* 2007, 24, 1415-1426.

[204] Joshi, HM; Bhumkar, DR; Joshi, K; Pokharkar, V; Sastry, M. Gold nanopartncles as carriers for efficient transmucosal insulin delivery. *Langmuir.* 2006, 22, 300-305.

[205] Tomii, A; Masugi, F. Production of anti-platelet-activating factor antibodies by the use of colloidal gold as carrier. *Jpn. J. Med. Sci. Biol.* 1991, 44, 75-80.

[206] Gibson, JD; Khanal, BP; Zubarev, ER. Paclitaxel-Functionalized Gold Nanoparticles. *Journal of the American Chemical Society.* 2007, 129, 11653-11661.

[207] Han, G; Ghosh, P; Rotello, VM. Functionalized gold nanoparticles for drug delivery. *Nanomedicine.* 2007, 2, 113-123.

[208] Ghosh, P; Han, G; De, M; Kim, CK; Rotello, VM. Gold nanoparticles in delivery applications. *Adv. Drug Deliv. Rev.* 2008, 60, 1307-1315.

[209] Fay, F; Scott, CJ; McCarron, PA. Recent Innovations in Antibody-Mediated, Targeted Particulate Nanotechnology and Implications for Advanced Visualisation and Drug Delivery. *Curr. Nanosci.* 2010, 6, 560-570.

[210] Reum, N; Fink-Straube, C; Klein, T; Hartmann, RW; Lehr, CM; Schneider, M. Multilayer Coating of Gold Nanoparticles with Drug-Polymer Coadsorbates. *Langmuir.* 2010, 26, 16901-16908.

[211] Zabielska, K; Lechowski, R. Novel approaches to cancer therapy. The use of nanotechnologies. *Med. Weter.* 2011, 67, 34-37.

[212] Hillyer, JF; Albrecht, RM. Gastrointestinal persorption and tissue distribution of differently sized colloidal gold nanoparticles. *J. Pharm. Sci.* 2001, 90, 1927-1936.

[213] Dambinova, SA; Izykenova, GA; Bogushev, BG, Drug addiction-revealing diagnostic reagent kit "narco-ifa-peptide test", Patent RU2297634-C2. RU2297634-C2 20 Apr 2007 G01N-033/94 200766.

[214] Selvaraj, V; Alagar, M; Hamerton, I. Analytical detection and biological assay of antileukemic drug using gold nanoparticles. *Electrochim. Acta.* 2006, 52, 1152-1160.

[215] Shen, J; He, F; Wan, Y; Feng, C; Zhao, Z; Wu, X; Zhu, L; Zhu, Y, New colloidal gold test paper card, comprises a reaction membrane, a junction releasing mat, and a testing region of coating quinolones drug-carrier protein coupling matter, useful for detecting quinolones drug residue, Patent CN101173925-A. CN101173925-A 07 May 2008 G01N-033/577 200852.

[216] Bi, D; Wang, X; Li, K; Xiong, N; Shi, D; Liu, B; Peng, F; Hao, W, Immuno colloidal gold test paper strip for detecting sulfa drug residue, Patent CN1888056-A; CN100396774-C. CN1888056-A 03 Jan 2007 C12N-005/18 200754 Chinese CN100396774-C 25 Jun 2008 200864.

[217] Rosi, NL; Giljohann, DA; Thaxton, CS; Lytton-Jean, AKR; Han, MS; Mirkin, CA. Oligonucleotidemodified gold nanoparticles for infracellular gene regulation. *Science.* 2006, 312, 1027-1030.

[218] Giljohann, DA; Seferos, DS; Prigodich, AE; Patel, PC; Mirkin, CA. Gene regulation with polyvalent siRNA nanoparticle conjugates. *Journal of the American Chemical Society.* 2009, 131, 2072-2073.

[219] Whitehead, KA; Langer, R; Anderson, DG. Knocking down barriers: advances in siRNA delivery. *Nature Reviews Drug Discovery.* 2009, 8, 129-138.

[220] Roemer, RB. Engineering aspects of hyperthermia therapy. *Annual Review of Biomedical Engineering.* 1999, 1, 347-376.

[221] Cherukuri, P; Glazer, ES; Curley, SA. Targeted hyperthermia using metal nanoparticles. *Adv. Drug Deliv. Rev.* 2010, 62, 339-345.

[222] Huang, W; Tsai, P; Chen, Y. Functional gold nanoparticles as photothermal agents for selective-killing of pathogenic bacteria. *Nanomedicine.* 2007, 2, 777-787.

[223] Huang, X; Jain, PK; El-Sayed, IH; El-Sayed, MA. Plasmonic photothermal therapy (PPTT) using gold nanoparticles. *Lasers in Medical Science.* 2008, 23, 217-228.

[224] Terentyuk, GS; Maslyakova, GN; Suleymanova, LV. Laser-induced tissue hyperthermia mediated by gold nanoparticles: toward cancer phototherapy. *Journal of Biomedical Optics.* 2009, 14, Article ID 021016.

[225] Sirotkina, MA; Elagin, VV; Shirmanova, MV; Bugrova, ML; Snopova, LB; Kamensky, VA; Nadtochenko, VA; Denisov, NN; Zagaynova, EV.

OCT guided laser hyperthermia with passively tumor-targeted gold nanoparticles. *Journal of Biophotonics.* 2010, 3, 718-727.

[226] Cardinal, J; Klune, J; Chory, E; Jeyabalan, G; Kanzius, J; Nalesnik, M; Geller, D. Non invasive radiofrequency ablation of cancer targeted by gold nanoparticles. *Surgery.* 2008, 144, 125-132.

[227] Hainfeld, JF; Slatkin, DN; Smilowitz, HM. The use of gold nanoparticles to enhance radiotherapy in mice. *Physics in Medicine and Biology.* 2004, 49, N309-N315.

[228] Kong, T; Zeng, J; Wang, X; Yang, X; Yang, J; McQuarrie, S; McEwan, A; Roa, W; Chen, J; Xing, JZ. Enhancement of radiation cytotoxicity in breast-cancer cells by localized attachment of gold nanoparticles. *Small.* 2008, 4, 1537-1543.

[229] Benford, ME; Chou, IH; Beier, HT; Wang, M; Kameoka, J; Good, TA; Cote, GL. In vitro detection of beta amyloid exploiting surface enhanced Raman scattering (SERS) using a nanofluidic biosensor - art. no. 68690W. In: VoDinh, T, Lakowicz, JR, editors. *Plasmonics in Biology and Medicine* V. vol. 6869. Bellingham: Spie-Int Soc Optical Engineering; 2008. p. W8690-W8690.

[230] Yoshiike, Y; Akagi, T; Takashima, A. Surface structure of amyloid-beta fibrils contributes to cytotoxicity. *Biochemistry.* 2007, 46, 9805-9812.

[231] Gold is newest weapon in battle against Alzheimer's. *Health News.* 2006, 12, 10.

[232] Takahashi, T; Tada, K; Mihara, H. RNA aptamers selected against amyloid beta-peptide (A beta) inhibit the aggregation of A beta. *Mol. Biosyst.* 2009, 5, 986-991.

[233] Yokoyama, K; Cho, HN; Cullen, SP; Kowalik, M; Briglio, NM; Hoops, HJ; Zhao, ZY; Carpenter, MA. Microscopic Investigation of Reversible Nanoscale Surface Size Dependent Protein Conjugation. *Int. J. Mol. Sci.* 2009, 10, 2348-2366.

[234] Yokoyama, K; Gaulin, NB; Cho, H; Briglio, NM. Temperature Dependence of Conjugation of Amyloid Beta Protein on the Surfaces of Gold Colloidal Nanoparticles. J. Phys. Chem. A. 2010, 114, 1521-1528.

[235] Pissuwan, D; Valenzuela, SM; Cortie, MB. Therapeutic possibilities of plasmonically heated gold nanoparticles. *Trends in Biotechnology.* 2006, 24, 62-67.

[236] Danscher, G; Larsen, A. Effects of dissolucytotic gold ions on recovering brain lesions. *Histochem. Cell Biol.* 2010, 133, 367-373.

In: Gold Nanoparticles ISBN 978-1-62257-927-3
Editors: A. Jarnagin and L. Halshauser

Chapter 2

GOLD NANOPARTICLES: FABRICATION BY LASER ABLATION TECHNIQUE AND THEIR OPTICAL AND MORPHOLOGICAL STUDIES

A. F. M. Y. Haider*
Physics Department. Faculty of Science,
University of Dhaka, Dhaka, Bangladesh

ABSTRACT

Material particles of the size of few nano-meters to couple of hundreds of nano-meters exhibit size and shape dependent properties and the consequent versatility and importance in their use have made them an exciting and interesting field of research. Of these, noble metal nano-particles, particularly gold nano-particles have shown tremendous potential for their wide range of applicability not only in fundamental sciences like physics, chemistry and biology, but also in applied sciences such as in designing and developing varieties of nano-particle based devices and sensors. The number of articles published on noble metal nano-particles, particularly gold and silver nano-particles has gone up in exponential rate since the early nineties of the last century.

One of the major thrust areas of nano-particles and their science and technology research are their fabrication. In the nineties of the last century and the first half of the last decade chemical methods dominated over all other methods, including the laser ablation technique which

* Email address: yuhaider@univdhaka.edu; yuhaider@gmail.com.

started emerging. The main disadvantage including a host of other drawbacks of conventional nano-particle generation techniques such as mechanical milling and grinding or even the chemical method like the sol-gel process is that it almost invariably picks up impurities which make them unusable and hence undesirable for many applications e.g. biological applications.

Pulsed laser ablation is an attractive and a promising alternative, addressing the shortcomings of the conventional methods, including the chemical methods.

In this chapter we discuss mostly about the nano-particle fabrication, with particular emphasis on gold nano-particles in colloids, using nano second laser pulses. A number of variables of the ablating laser pulse have been used to control the size of the fabricated nano-particles. These were done without the use of chemicals/chemical surfactants, which provided a good and clear understanding of the laser parameters on the control of the nano-particle size. Possible explanations of all observations are given.

Introduction

Nanotechnology or 'Nano science' involves the ability to fabricate, characterize and manipulate structures at nanometer scale. It is one of the greatest steps forward in scientific knowledge and technological advancement of the new millennium. This new technological revolution is going to introduce nanotechnology in all aspects of our daily life and make pathways for advancement of inexpensive but (energy) efficient technology and health care system which will certainly bring new economic possibilities for the world.

Nanoparticles are aggregate of atoms (molecules) in the nanometer scale, whose properties are neither those of the bulk, nor the individual atoms (or molecules). Properties of these nanoparticles depend more on the size (and to some extent the shape) rather than the material itself, resulting in reliable and gradual change in any property that can be obtained using a single material [1]. This is because as the wave function of the electron is confined within a small volume of the particle, the energy levels of the electrons take discrete values (instead of bands) and results in diverse (changes in) physical properties e. g. in quantum dots, which are nanometer scale semiconducting entities.

For example as the size of quantum dots becomes smaller and the wave function is confined within a smaller volume, the energy of the electrons increases in accordance with Heisenberg's uncertainty principle. This is called the "quantum size effect." As a result, with quantum dots even of the same material, it is possible to change the wavelength of the emitted light (i.e., color), which depends on the size of the quantum dot. Also because the luminescence peak is sharp, large optical amplification at the peak wavelength is possible.

The Coulomb repulsion between electrons makes the probability of multiple free electrons existing in a single quantum dot small, which means that light emission and electrical conduction are realized by one electron. It is also possible to observe the spin of single electrons. The Coulomb attraction between electrons and holes is large, and hence the resulting excitons, are stabilized more in quantum dots.

Again since de-phasing of the electrons is small, quantum dots are advantageous for realizing quantum computation, in which a long de phasing time is necessary. Furthermore, in comparison with organic dyes, quantum dots also have the advantage of minimal deterioration due to repeated light absorption.

However there are difference in the origin of the size dependent properties of noble nanoparticles and that of semiconducting nanoparticles. Colloids of gold or silver nanoparticles in water show size dependent extinction peaks (hence colors) in the visible range of the electromagnetic spectrum. This property is attributed to the collective oscillation of the electrons, in the nanoparticles, and is known as the surface plasmon oscillation.

Because of the size (and shape) dependent properties of the nanoparticles they have a wide range of applications not only in fundamental sciences like physics, chemistry and biology but also in the applied sciences such as in designing and developing a variety of electronic devices (molecular electronics). In electronics, the design and the assembly of functional materials and devices based on nanoparticles can be seen as the natural inevitable evolution of the trend toward miniaturization. Silicon based electronics is reaching its limits to achieve such feat. Therefore, nano-tubes and quantum dots are becoming promising candidates for building nano-transistors. This will offer higher device versatility, faster switching speed, lower power dissipation and possibility of packing many more transistors in a single chip.

An important application of nano particles could be in biological systems. The pharmaceutical and biomedical industries are trying to synthesize supra-molecular assemblies and artificial devices which will mimic the complex mechanism of natural process. This can be used for more efficient diagnosis and better cures for diseases. For example using porous nanoparticles and nano-capsules such as liposomes, drugs can be selectively released in living organs using the 'size effect' of nanoparticles or bio-conjugated assemblies of bio-molecules. This will lead to faster and more selective drug design and drug delivery [2].

Other important examples of nanotechnology applications are: a) giant magneto-resistance in nano crystalline materials which help to build smaller information storage devices, b) nano-layers with selective optical barriers, c) chemical and bio-detectors, d) advanced drug delivery systems, e) chemico-mechanical polishing with nanoparticle slurries, f) nano-crystals in generating novel laser and electrical devices and g) carbon nano-tube products etc.

The most challenging aspect of nanotechnology is the production of nanoparticles of desired size (and shape) and a number of refined chemical synthetic methods have evolved in the recent past for this purpose [1].

Nano particle fabrication can be broadly classified in two groups viz. the chemical method and the physical method. In chemical method, there are three basic techniques for fabrication of nano-particles. They are: vapor-phase synthesis, solution precipitation and solid-state process. Nanoparticles can be generated either by adopting any of these processes solely or from their various combinations. In the early era of nanoparticle generation, vapor phase processes had been used widely but in recent time solid state process have been adopted extensively for generation of nanoparticles as they are more cost effective process. In general, conventional chemical methods for fabrication of nanoparticles are:

- Solid-State Synthesis
- Vapor-Phase Synthesis
- Sol-Gel Processing
- Solution Precipitation
- Water in oil Micro emulsion (Reverse Micelle) Method

The physical methods, in general, offer the possibility of nanoparticle growth in a more controllable, contamination free environment. Here is a mention of commonly adopted physical methods along with a focus on laser ablation technique.

- Mechanical milling and grinding
- Inert gas condensation
- Arc discharge
- Ion sputtering
- Laser ablation

One can also broadly classify the techniques as bottom up or top down depending on whether the nano particles grow up or breaks down to the desired size. Surfactants, even though they are not involved in the chemical process of nano structure fabrication, play a significant role in their fabrication. The mechanism of the role of the surfactants has been discussed by a number of researchers [3-6]. Further study is necessary to ascertain the exact mechanism of the involvement of the surfactants, in the fabrication process which is almost invariably used in chemical process. However, it is believed that the surfactants help in preventing coagulation of nanoparticle in colloids by producing a protective coating around the nano particles.

Conventional chemical methods for fabrication of nanoparticles suffers from drawbacks like impurity pick up and needs the use of precursor to remove chemical by-products. However this does not ensure the total removal of the chemical debris rendering the nanoparticles produced by chemical means unsuitable and hence unusable where purity is a prime concern e. g. in biological applications.

On the other hand, pulsed laser ablation technique is an attractive and promising alternative which removes all these problems of chemical methods bypassing them. This is a rapid, simple and most importantly the purest method for preparation of nanoparticles particularly in colloids.

In pulsed laser ablation technique laser energy is used to remove a portion of a sample by melting, fusion, sublimation, ionization, erosion, and/or explosion. At the very beginning, plasma is formed above the surface of the material. Later when the plasma cools down, nanoparticles are produced by the condensation process. The following diagram shows the schematic shape of a laser induced plasma plume.

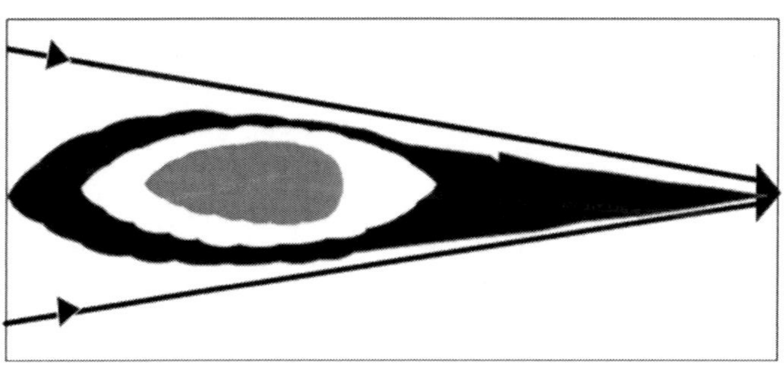

The laser-induced plasma undergoes a pear or cigar-shaped non-isotropic expansion from the focus of the laser beam having high electron and ion densities (10^{17} - 10^{20} cm^{-3}) with temperature in the range of 5000-20000 K.

With nano second pulsed lasers the power requirement for the production of plasma is usually in the range of $10^7 – 10^{11}$ W/cm^2 [7].

The interest in gold nanoparticles has significantly revived with better understanding of quantum mechanical nature of nanoparticles, nature of interaction between colloidal particles and dependence of colloidal stability against coagulation in presence of surfactants. The corresponding advances in technology related to optical and electron microscopy, optics, lasers, computing facilities for data acquisition and analysis and instruments designed with better precision encouraged the research, in gold nanoparticle in particular, and nano science and technology in general, for useful applications in better and smaller chip design, cancer treatment and targeted drug delivery systems.

Gold nanoparticles are promising probes for biomedical applications because they can be easily prepared and unlike other fluorescent probes such as quantum dots and organic dyes do not burn out after long exposure to light. It can be used to identify biomarkers in the very early diagnosis of liver, lung and breast cancer.

A technological breakthrough is in the offing with the next generation computers using hybrid of organic molecules and gold nanoparticles. Moreover, the latest gold rush is likely to revolutionize the field of bio-sensing and chemical sensing by using highly environmentally sensitive nature of plasmon resonance as well as other optical effects (e. g. surface enhanced Raman or fluorescence) associated with not only gold but also with other noble metal nano particles .

Pulsed laser ablation in liquid media permits synthesis of gold nano particles with fair control of size [8-11]. Optical extinction spectroscopy is used to determine the size of spherical gold nano particles suspended in solution [12, 13] with the extinction spectra generally undergoing a blue shift with the reduction of nano particle size [14]. In pulsed laser ablation the reduction in nano-particle size can be effected by increasing laser pulse energy. However, Simakin et al. [15] did not observe any apparent red or blue shift of the extinction peak of the gold colloids with increasing ablating laser fluence of 10-20 J/cm^2. Kazakevich et al. [16] observed a non-monotonic shift of the extinction peak with laser fluence, their maximum fluence being 35 J/cm^2. Again, Baracikowski [17] reported an increase of the mean size of the generated nano-particles at higher laser pulse energy. They used femtosecond laser pulses of energies 150 μJ and 300 μJ.

Pulsed laser ablation process in liquids involves the superposition of two mechanisms, viz., (a) the ablation of the material and (b) the interaction between the laser and the ablated nano particles. Riabinina et. al. [18] studied the influence of laser irradiance alone on gold colloids prepared independently by chemical methods. They used KrF Excimer laser (wavelength of 248 nm, pulse width of 17 nsec, repetition rate of 20 Hz and pulse energy of 50 mJ). However their study became complicated with the physico-chemical process induced by laser-matter interaction particularly due to the residual chemicals present in the gold colloids [19].

Although nanotechnology and nanoparticle research hold enormous benefits, they are still far from commercialization because of the need for extremely sophisticated fabrication method and tools to manipulate at nano-scale. But recent advances in developing self assembly of nano-particles and also the development of efficient but cheap fabrication based on laser ablation and other techniques have accelerated the activities in this sector much greater than before. USA, Japan, China, South Korea, India, Brazil and EU have already invested huge amount of money to achieve nanotechnology at the industrial level [2].

It is essential that developing countries, despite of their resource constraints, should also take part in research of nano-technology so that they can reap the benefits of its capabilities. Nanoparticle research, can address typical problems of developing countries, like effective water treatment to check water pollution, arsenic poisoning and water-borne diseases like cholera and diarrhea. Also it may help to develop environment friendly manufacturing of cheap solar cells and biomedical sensors [2].

In this review paper we present the results of our experiment on the formation of gold colloids over a wide range of laser fluence without the use of any chemical/surfactant [19]. The effect of re-ablation time on freshly prepared gold colloids by laser ablation was also studied. We have also studied the dependence of size of the nanoparticles on laser ablation energy by establishing correlation between them. These correlations can enable one to generate nanoparticles of desired size by suitably adjusting the laser ablation parameters. The nanoparticles of desired size, thus produced, can be utilized to monitor water pollution, arsenic poisoning in various water-treatment plants, monitor water-borne diseases like Cholera and Diarrhea and also in manufacturing cheap sensors to detect adulteration especially in edible substances [19].

Before presenting our experimental details let us review the experimental and theoretical study related to the formation of metal nano-particles by laser ablation technique and their consequent optical and morphological characteristics.

Theory of Nano-Particle Generation by Laser Ablation

This section presents a description of theoretical background and advantages of nanoparticle generation by laser ablation and also illustrates various consequences of optical properties of metal nano-particles emanating from surface plasmon resonance.

Brech and Cross [20] in year 1962 made the first reported study of the plasma spectrum of the vaporized solid using ruby laser. This work was the basis for the technique of laser microprobe emission spectroscopy for the elemental analysis of solid materials. Throughout the 60's, extensive studies on laser ablation technique had been carried out. Linlor [21] measured the energy of the ejected ions in the laser-induced plasma. Honig and Woolston's work [22] showed the first example of laser AMSS spectrograph which was commercially instrumented by Leybold-Heraeus in 1978. Another significant study was done by Rosan et al. [23] on the ablation of biological materials. Berkowitz and Chupka [24] were the first to observe the clusters in ablation plume. The advanced laser technology enabled Zaitsev-Zotov et al. [25] to produce superconducting thin film in 1983. Since 1980 there has been a tremendous spur in laser ablation research activities by developing various applications in medicine, biology and industry.

The production of nanoparticles by laser ablation technique is a relatively new concept in nano-material fabrication process. Formation of metal clusters was reported at first by Dietz et al. [26] in 1981. In the following years, vaporization by laser and formation of superfine particles by cooling it in gas or by supersonic expansion were reported [27, 28]. The first production of metal nano-particles was reported by Becker et al. [28] in 1998. Since then there has been extensive theoretical and experimental studies of factors which influence the fabrication, particularly in controlling the size, stability and functionality of both organic and inorganic nano-particles by thermal and non-thermal ablation. The field came in the forefront of research with the aim of developing controlled novel nano-particle fabrication techniques.

Laser ablation involves several stages which are governed by a variety of distinct nonlinear mechanisms. Once the laser beam illuminates the sample, a small mass is ablated from the surface of the sample in the form of electrons, ions, atoms, molecules, clusters, and particles, each of the processes separated in both time and space. Nanoparticle fabrication by laser ablation can be divided into three main processes: bond breaking and plasma ignition, plasma expansion and cooling, and particle ejection and condensation. These laser ablation processes occur over several orders of magnitude in time, starting with electronic absorption of laser optical energy from 10^{-15} sec to particle condensation at 10^{-3} sec after the laser pulse is completed. Russo et al. [7] wrote an excellent chapter on "Laser ablation" in a book titled "Laser Induced Breakdown Spectroscopy" edited by P Singh. Following Russo et al. [7] a brief description of different stages of laser ablation technique is presented here.

The primary ionization in laser-induced plasma is initiated by multi-photon absorption of the focused laser light. For a nanosecond laser pulse with irradiances around 10^8 W/cm^2, the dominant mechanism is thermal vaporization. The laser strikes the material surface and gets absorbed by the atoms/molecules and primary electrons in the solid. This absorption leads to the high temperature of the ablated solid, and a well defined phase transition occurs, from solid to liquid, liquid to vapor, and vapor to plasma. For a picoseconds laser pulse with irradiance between 10^{10}–10^{13} W/cm^2, both thermal and non-thermal mechanisms exist. In a non-thermal mechanism, where Coulomb explosion process dominates, electronic excitation energy due to the absorption of intense laser leads to energetic atomic motion that breaks the bonds in the solid and forms charged fragments which get separated from each other due to coulomb repulsion among them. For irradiances higher than 10^{13} W/cm^2 with femtosecond laser pulse, Coulomb explosion is the main bond breaking mechanism.

When the laser pulse duration is in the nanosecond time region, the later part of laser pulse can be absorbed by the laser induced plasma. For picoseconds pulsed laser ablation, the laser pulse is too short to be absorbed by the plasma. In the plasma plume the laser is primarily absorbed by inverse Bremsstrahlung processes and direct single-photon processes [29]. Inverse Bremsstrahlung involves absorption of photons by free electrons. The other contribution comes from multi-photon ionization process which becomes more dominant if the laser is of shorter wavelength. These two processes determine how much ionization degree can be achieved in the initial weakly ionized plasma plume and thus the amount of solid mass converting into vapor and the behavior of the resulting vapor.

The plasma expansion process depends on number factors. These are the amount and properties of the ablated mass, the energy coupled into the mass, spot size of the laser beam and the environment. For a laser with fluence of 1-10 J/cm^2 on a laser spot size of 0.01-0.1 cm^2, the no. of ablated particles from a metal is in the range of 10^{15} to 10^{16} atoms/pulse. This leads to a typical plume density of 10^{19}-10^{20} atom/cm^3. In the vacuum the expansion of plasma plume is adiabatic. In gaseous or liquid environment the ejected mass compresses the surrounding media and thus produces shockwaves [7].

Nano-sized particles are formed from condensation of the vapor. Condensation starts when the vapor plume temperature reaches the boiling temperature of the material (~3000 K) and stops at the condensation temperature of the material (< 2000 K).

Advantages of Laser Ablation

Nanoparticle generation by methods like mechanical milling and grinding, or chemical generation methods like that of sol-gel process show drawbacks related to the purity as mentioned earlier. The use of precursor is often required during the production process, so that purification steps to remove the chemical by-products have to be carried out, which are quite complicated in colloids, particularly, with reactive metal nanoparticles. Gas phase processes give powders, which are often hardly re-dispersible into liquids or almost impossible to grind down to the primary particles.

On the other hand pulsed laser ablation, as an alternative method, addresses the problems of the methods mentioned above. By this technique nanoparticles may be generated from almost any solid material. Nano-particles fabricated by this method have high purity of the nano-material; have a large material variety for nano-particle fabrication, with in-situ dispersion capability of the nanoparticles in a variety of liquids which allows safe and stable handling of the colloids.

It may also be possible to achieve the intended nanoparticle size by varying laser pulse duration and pulse energy [19, 30]. It also reduces the contamination risk. Of late, various forms of complex oxide films, quantum dots and one-dimensional nanostructures, such as nano-wires, nano-tubes and other composite nanostructures have been fabricated by this technique.

Optical Properties of Noble Metal Nanoparticles

In this section the optical properties of gold nanoparticles, in particular, and their response to external electromagnetic field which gives rise to the localized surface plasmon resonance (SPR) and its (SPR) dependence on the particle size, incident electromagnetic wave and the environment, will be discussed in brief. We discuss here how the electronic properties of the nanoparticles lead to the size and shape dependent absorbance of gold colloids.

In the 1850s Michael Faraday [31] observed the transmission of green light by extremely thin gold foil. Faraday supposed that a material that is capable of strong interaction with light can do so when its dimensions became smaller than visible wavelengths. He synthesized the characteristic ruby-red Au particle suspensions through a number of methods and showed that while the suspensions reacted to acids in the same way as bulk gold, the particles were apparently too small to be resolved by his microscope. Faraday also demonstrated that the color of gold colloids would predictably change when varying amounts of aggregation-inducing agents were added in the colloid.

A classic work was done by Gustav Mie in 1908 [32] who published his work on the extinction of light by metal spheres. Starting from the Maxwell's equations, Mie calculated the extinction, scattering, and absorption cross-sections of metal sphere of arbitrary size and also showed how the spectra of the metal particle suspensions evolve as a function of particle size. The results of these calculations also allowed him to sketch scattering diagrams for different particle sizes and diagrams depicting the electric and magnetic fields of the dipole, quadruple, octupole, and sextuple components of the resonance. But applicability of Mie's theory was restricted to only spherical nanoparticles. So when he compared his calculation with the experimental results of his doctoral student W. Steubing as well as with the results of Faraday and Zsigmondy [31, 33], he came across the deviation from his theoretical result which he attributed to the non-spherical size of the nanoparticles. Later Gans [34] extended the theory to consider spheroid shaped particles showing how smaller, aspherical particles can show absorbance of longer wavelengths than the spherical one of comparable size. At present day, fast computers along with various numerical methods have been used to analyze the response of ellipsoidal, rod-like and cubic shaped nanoparticles. [35].

Extinction Spectra

Scattering, both Rayleigh and Mie, and absorption are the main interactions of light with matter containing particles both of which cause attenuation in the intensity of the beam passing through the medium. Scattering is caused by heterogeneity in the medium caused simply by density or concentration fluctuations, or by presence of particles or drops as dispersed phase. When one deals with particles dispersed in a fluid or a gas, typically the scattering by fluctuations in the media is much lower than by the particles and can be neglected for practical purpose. The blue color of the sky is a result of Rayleigh scattering by the air molecules, and the attenuation of light passing through turbid media like fog is also due to scattering. Again, scattering is often accompanied by absorption, whereby depending upon the material in question, certain frequencies are absorbed and dissipated either as heat or other internal processes in the molecules or particles. The color of black smoke or of Indian ink dispersed in water is due to absorption. The combined effect of scattering and absorption is called extinction.

A light beam passing through a colloidal dispersion of metal nanoparticles gets attenuated by the combined contribution of absorption and scattering, and is given by

$$I(x) = I_0 \exp(-N_0 C_{ext} x) \quad (1)$$

Here, I_0 is the intensity of the incident beam. $I(x)$ is the intensity of the beam after travelling distance x within the sample, N_0 is the number density of the particles and C_{ext} is the extinction cross section of a single particle and is the sum of the absorption cross section C_{abs}, and scattering cross section C_{sca}. The term N_0C_{ext}/a, where a is the area of the particle, is often referred to as the extinction coefficient, Q_{ext}. The absorption spectrum (observed by UV-Vis spectroscopy) is in fact a measure of attenuation caused by dispersion of nanoparticles, and is thus related to the absorption and scattering cross section.

The relationship between the size (and shape) of nano-particles, in particular of the gold nano-particles, and the extinction cross sections establishes the physical basis for their optical response. This can be obtained from Mie's theory and classical electromagnetic theory [36, 37 and 38].

Absorption and Scattering by a Single Particle

The treatment of absorption and scattering of light by small particles is a problem of electromagnetic theory which is governed by the Maxwell's equation with proper boundary conditions. When a particle is illuminated by a beam of light with specified characteristics, the intensity and angular distribution of the light scattered by the particle, as well as the amount absorbed, depends in a detailed way on the nature of the particle, that is, its shape, size, and the materials of which it is composed. This presents us with an almost unlimited number of distinct possibilities. Nevertheless, there are some features common to the phenomena of scattering and absorption by small particles. So the problem can be formulated in the following way:

Given a particle of specified size, shape and optical properties that is illuminated by an arbitrarily polarized monochromatic wave, one has to determine the electromagnetic field at all points in the particle and at all points of the homogenous medium with known dielectric constant, in which the particle is embedded.

The exact solution for such problem has been given by Mie in 1908 [32] for the case of a sphere of arbitrary radius and refractive index. Detailed calculations are also available in literature [39, 40, 41].

Plasmon in Gold Nanoparticles

Let us consider the dielectric function of the electron gas $\varepsilon(\omega)$ in the nanoparticle which can be written in term of its real and imaginary parts as

$$\varepsilon(\omega) = \varepsilon'(\omega) + i\varepsilon''(\omega) \quad . \tag{2}$$

Drude's free electron model, despite its limitations in describing various phenomena of metals, can be used to understand some optical response of the metal. According to this model, outer shell electrons are free and act as a gaseous substance. So one can consider metal as medium with equal concentration of positive ion cores and negative mobile electrons i.e. plasma. The associated plasma frequency ω_p is given by the following equation [42].

$$\omega^2{}_p = \frac{N_o e^2}{m^* \varepsilon_o} \tag{3}$$

Where N_o the density of the free electrons, m^* is the effective mass of the electron and ε_o is the electric permittivity of vacuum.

The dielectric function for free electron gas is given from Lorentz formula:

$$\varepsilon = 1 - \frac{\omega^2{}_p}{\omega^2 + i\gamma\omega} \quad , \tag{4}$$

with real and imaginary parts,

$$\varepsilon' = 1 - \frac{\omega^2{}_p}{\omega^2 + \gamma^2} \text{ and } \qquad \varepsilon'' = \frac{\omega^2{}_p \gamma}{\omega(\omega^2 + \gamma^2)} \quad , \tag{5}$$

ω is the angular frequency of the incident e. m. wave and γ is the damping constant respectively (discussed later).

If the electrons in the metal are disturbed slightly from the equilibrium by some means, the non-uniform charge distribution will set up an electric field that will tend to restore charge neutrality. The electrons, having acquired momentum will overshoot the equilibrium and there will be oscillation. This oscillation is longitudinal and is due to the collective oscillation of the electrons in the metal. This oscillation is known as plasma oscillation. The quantum of such oscillation is called bulk plasmon.

Surface Plasmon

The optical properties of noble metal nanoparticles originate from localized surface plasmons which are different from usual bulk plasmon observed in the metal. These phenomena occur when electromagnetic field interacts with conduction band electrons and induces the coherent oscillation of electrons. As a result, a strong absorption band appears in some region of the electromagnetic spectrum depending on the size of the particle. This plasmon absorption is a small particle effect. It is absent in the individual atoms as well as in the bulk.

But when a metal nanoparticle is exposed to such a field, non propagating excitations of free electrons create size dependent localized surface plasmons that arise simply when we consider the absorption and scattering by gold or other metallic nanoparticles. When a conductor or metal is placed in an oscillating field of incoming radiation, the electrons cloud is driven into oscillations.

In the case of a sub-wavelength conductive nanoparticle, the curved surface of the particle exerts an effective restoring force on these driven electrons (analogous to a damped and forced harmonic oscillator). Like any driven oscillator system, in nanoparticle case, a resonance can arise leading to the field amplification both inside and outside the particle. This resonance for gold and silver particles lies in the visible region of the electromagnetic radiation which is responsible for their bright colors in both transmitted and reflected light [19, 30]. In the following section we will show the size dependent properties of spherical nanoparticles from Mie's theory of scattering and absorption.

Localized Surface Plasmon Resonance for Spherical Particles

As the radius of the particle becomes comparable to the wavelength of the incident electromagnetic wave (ranging from UV to visible light) one can write down the absorption and scattering coefficients, by considering the radius of the sphere 'a' as a much smaller quantity than 'λ' i.e. by assuming x << 1 and also |m|x << 1, as the following [40, 41]:

$$Q_{abs} = 4x\,\mathrm{Im}\left\{\frac{m^2-1}{m^2+2}\right\} \tag{6}$$

and

$$Q_{sca} = \frac{8}{3}x^4\left|\frac{m^2-1}{m^2+2}\right|^2, \tag{7}$$

Where x and m are defined as, $x = ka = \frac{2\pi a}{\lambda}$ and $m = \frac{k_l}{k} = \frac{n_l}{n} = \sqrt{\frac{\varepsilon_l}{\varepsilon_m}}$ for $\mu = \mu_l$, and ε_l and ε_m are the permittivity of sphere and surrounding medium respectively.

Here n and n_l are the refractive indices of the medium and the particle respectively.

k and k_l are the propagation constants in the medium and the particle respectively and the μ's are the corresponding permeabilities.

Putting the value of m one can find for a small sphere:

$$Q_{abs} = 4x \operatorname{Im}\left\{ \frac{\varepsilon_l - \varepsilon_m}{\varepsilon_l + 2\varepsilon_m} \right\} \tag{8}$$

and

$$Q_{sca} = \frac{8}{3} x^4 \left| \frac{\varepsilon_l - \varepsilon_m}{\varepsilon_l + 2\varepsilon_m} \right|^2 \tag{9}$$

We use electrostatic approximations and consider only the dipole moment interaction with the incident electromagnetic field for the time being as it is the dominant interaction mechanism with the nanoparticle and the electromagnetic field.

As the dipole moment is oscillating because of the oscillating electric field of the incident e. m. wave, the scattering and absorption cross sections from coefficients given in the eq. (8) and (9) can be written (in terms of cross section C, given by C = Coefficient (Q) × Area (the e. m. wave sees an area of πa^2 of the nanoparticle of radius a)) as:

$$C_{sca} = \frac{8\pi}{3} k^4 a^6 \left| \frac{\varepsilon_l - \varepsilon_m}{\varepsilon_l + 2\varepsilon_m} \right|^2 \tag{10}$$

$$C_{abs} = 4\pi k a^3 \operatorname{Im}\left\{ \frac{\varepsilon_l - \varepsilon_m}{\varepsilon_l + 2\varepsilon_m} \right\} \tag{11}$$

The applied field is inducing polarization in the sphere and the required energy to polarize, is provided by the incident wave. Maximum polarization will require maximum amount of energy to be absorbed by the sphere. So by the requirement of maximum value of polarizability, one can derive a resonance condition (as the spherical nanoparticle is now acting as a forced harmonic oscillator whose amplitude is proportional to the induced dipole moment, P) for a given spherical nanoparticle. Polarizability,

$$\alpha = 4\pi a^3 \frac{\varepsilon_l - \varepsilon_m}{\varepsilon_l + 2\varepsilon_m}$$

Now the maximum value is reached if $|\varepsilon_l + 2\varepsilon_m|$ is a minimum, which, for the case of small or slowly varying ε_l around the resonance condition, simplifies to [2, 40, and 41]

$$\mathrm{Re}\,\varepsilon_l = \varepsilon_l' = -2\varepsilon_m \tag{12}$$

This condition is known as the Frohlich condition and the frequency for which this relation is satisfied is called Frohlich frequency. The associated mode of oscillation is called the dipole surface plasmon of the nanoparticle, and for a given metal, depends on the permittivity of the medium as well.

In the plasma model, when the electrons oscillate in response to the applied electromagnetic field, their motion is damped via collisions among them that occur with the frequency of $1/\gamma$ where γ is the damping constant. Since electron–electron, electron–phonon and electron-defect scattering processes in the bulk material determine the value of γ, it remains constant for the bulk material. But when the particle size itself is comparable to the mean free path of the collision, collisions of the free electrons with particle surface also contribute to the damping constant. So the damping constant becomes,

$$\gamma(r) = \gamma_o + \frac{Av_F}{r} \tag{13}$$

where γ_o is the constant bulk damping constant, v_F is the velocity of free electrons at the Fermi energy and A is a parameter taking into account the contributions of all types of scattering processes i.e. electron-electron, electron-phonon, electron-defect scattering and electron- particle surface collision etc. occurring in the metal. This size dependence of the damping constant makes the permittivity function in eq. (5) and hence the resonance condition, eq. (12), a function of the nano-particle's radius, r (or a) [2, 40, and 41].

From eq. (8) and (9) for x=ka <<1, we can see that scattering coefficient Q_{sca} and hence the scattering cross section, C_{sca} is proportional to x^4 and is much less than absorption cross section, C_{abs} ~ x. Therefore in this limit absorption dominates over scattering and the extinction cross section is effectively the absorption cross section i.e.

$$C_{ext} = = C_{abs} + C_{sca} \sim C_{abs}\ .$$

Denoting local dielectric function ε_l inside the sphere as ε (dropping the subscript, 'l') the extinction cross section for spherical nanoparticle of radius 'a' embedded in the medium with permittivity ε_m can be written as [2, 40, and 41],

$$C_{ext} = \frac{24\pi^2 a^3 \varepsilon_m^{3/2}}{\lambda} \frac{\varepsilon''}{(\varepsilon' + 2\varepsilon_m)^2 + \varepsilon''^2}, \tag{14}$$

Where λ is the wavelength of incident electromagnetic wave and ε' and ε'' are the real and imaginary parts of the dielectric function of the nanoparticle. This relation clearly indicates the absorbance peak shift with varying size and giving rise to different color of the nanoparticle solution.

The Frohlich condition, eq. (12) derived above was for the limiting case of a particle with vanishing size. The Frohlich condition for the finite size but small particle is [45, 2],

$$\varepsilon' = \operatorname{Re}\varepsilon = -(2 + \frac{12}{5}x^2)\varepsilon_m \tag{15}$$

Now the real part of the permittivity function is always an increasing function of the frequency as can be seen from eq. (5) so an increase in particle size (as x=ka) shifts the Frohlich frequency to the lower value of frequency (i.e. to longer wavelength) so that the above relation is maintained. This implies that for Frohlick condition to be satisfied the ratio of a/λ should remain constant. So if 'a' increases, 'λ' the peak wavelength of the extinction spectra should also increase linearly.

From above discussion we can see that the size dependence of absorption by nanoparticles and the observed absorbance peak shifting towards lower or higher values of wavelength with decrease or increase of nanoparticle size are expressed through the relations in equations (13) to (15).

Gold Nanoparticle Fabrication

One of the crucial aspects of nanoparticle research is achieving novel fabrication technique which not only generates nanoparticles of desired sizes, but also renders them into various functional assemblies; which, in turn, can be applied in biological, material and industrial sectors. Methods like attrition, pyrolysis, evaporation and ablation by plasma, sol-gel processing are the frequently used fabrication techniques of nanoparticles. Each fabrication process has got its own pros and cons; for example, 5-10 nm sized nanoparticles can be produced by a relatively simple chemical reduction method but the surface of these nanoparticles is likely to be contaminated with reactions by-products such as anions and reducing agents, which can interfere with subsequent stabilization and functionalization steps. For this reason, much emphasis has been given on generation of nanoparticles by purely physical methods like laser ablation.

Laser Ablation Technique

The definition of ablation is usually referred to as "loss of a part by melting or vaporization". Ablation from a material occurs when energy is absorbed by it. This energy can be provided in the form of electrical discharges (e.g. an arc and spark) or in the form of light (e.g. as a laser). Laser ablation means using laser energy to remove a portion of a sample by melting, fusion, sublimation, ionization, erosion, and/or explosion as mentioned earlier. Laser ablation results into the formation of a gaseous vapor, luminous plasma, and in the production of fine particles of various sizes depending on several factors.

Experimental Setup for Nanoparticle Generation

Figure-1 shows the schematic diagram of the laser ablation experimental set up used for the fabrication of nano-particles of Gold (Au). Nanosecond light pulse (the 2nd harmonic at 532 nm) from a high power frequency doubled Q-switched Nd: YAG laser with a pulse width of ~8 ns and a pulse repetition rate of 10 Hz was focused by a lens of focal length 7.5 cm on to small circular area (150 μm diameter) of the sample (Au). When a laser pulse of irradiance higher than a certain threshold (generally> 0.9 GW/cm^2) is focused on the sample, it instantly generates a plasma plume, with temperature in the range of 8,000-12000K [43]. Nanoparticles are formed during condensation.

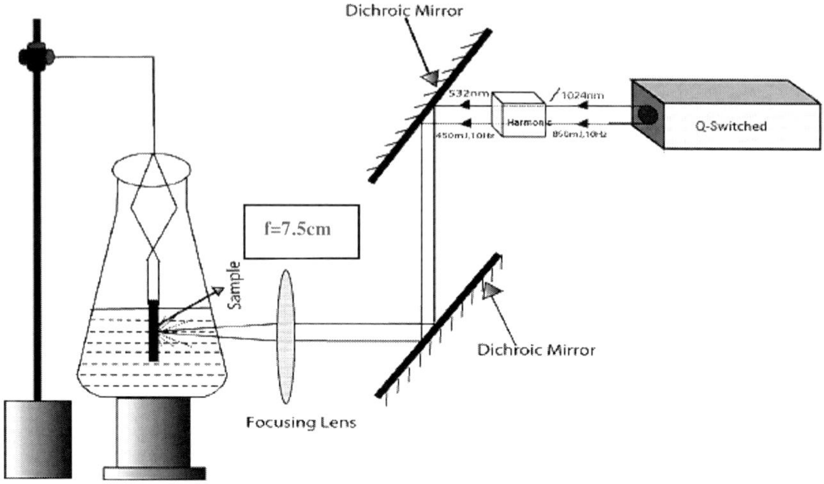

Figure 1. A schematic diagram of the experimental setup for laser ablation technique.

Formation of Nanoparticles

A 99.99% pure Suisse gold bar of weight 5g from Switzerland was used as bulk sample for generating nanoparticles of gold. For verification of its purity both LIBS and XRF measurements were performed for the sample. The gold bar was placed in a glass vessel with 30 ml of de-ionized nanopure water, such that water height from the top edge of the gold bar was about 5 mm, that is, the bar was well immersed in water from all sides. A photograph of the gold bar is shown in Figure 2 (a) and figure 2 (b) shows the produced colloid of gold nanoparticles in nanopure water with distinct color.

The preparation of metal nanoparticles appears simple, but it requires a great deal of care to have reproducible and stable colloids. Prior to the laser ablation, the gold bar and the ablation cells (glass vessel) were washed and cleaned in ultrasonic bath. Then the pulsed laser output (2nd harmonic of the Nd: YAG laser) was focused on the gold bar in water. The energy per pulse of the laser beam was controlled by the laser flash lamp discharge voltage and measured with an optical power meter.

- The position of the focus on the sample was adjusted until the maximum cracking sound was heard.
- An X-Y-Z translational stage was used, so that position of the sample can be adjusted in 3 dimensions in order to place the sample exactly at the focal point of the laser beam.

(a)

(b)

Figure 2. (a): The bulk gold sample. 2(b): Gold colloid generated from the bulk gold sample.

Characterization: Optical and Morphological

The optical (extinction spectra) and morphological study (particle size and shape) was done by using UV-Visible Spectroscopy and Scanning Electron Microscopy (SEM/EDX).

RESULT AND DISCUSSION

The eight ablation energies were used in the experiment to study the dependence of the size of the gold nanoparticles on laser pulse energy. Table 1 shows the Ablation energies along with the corresponding values of fluence and laser power.

Table 1. Ablation energies along with corresponding values of fluence and laser power

Observation no.	Ablation energy= Laser energy per pulse(mJ)	Laser wavelength (nm)	Pulse Width, τ (ns)	Spot size d (μm)	Laser power, P=E/τ MW	Fluence= $E/\pi(d/2)^2$ (Jcm^{-2})
1	18	532	8	150	2.25	102
2	34				4.25	192
3	52				6.50	293
4	73				9.13	412
5	96				12.0	542
6	122				15.25	689
7	150				18.75	847
8	184				23.0	1039

COLOR CHANGE

In the de-ionized nanopure water, the ablation led to common phenomena of change of color of the colloid. Within a short time after the ablation was started, the color of water turned into a violet-reddish tone. The color of Au colloid, obtained at laser pulse energy of 18 mJ, was medium violet red and it shifted from reddish to bluish tone with increase of laser pulse energy (figure 3). This well-known coloration of the solution has clearly indicated the generation of gold nanoparticles in water.

The absorption spectra of the colloidal solution were taken immediately after fabrication. Figure 4 represents the extinction spectra (which is mostly due to absorption) of the gold nanoparticles fabricated at different laser energies. The color of the colloid is due to the complementry reflected light. So with blue shift of the extinction peak with the laser pulse energy one observes a change in the color of gold colloid.

Figure 3. Change in coloration of gold colloid with increasing energy.

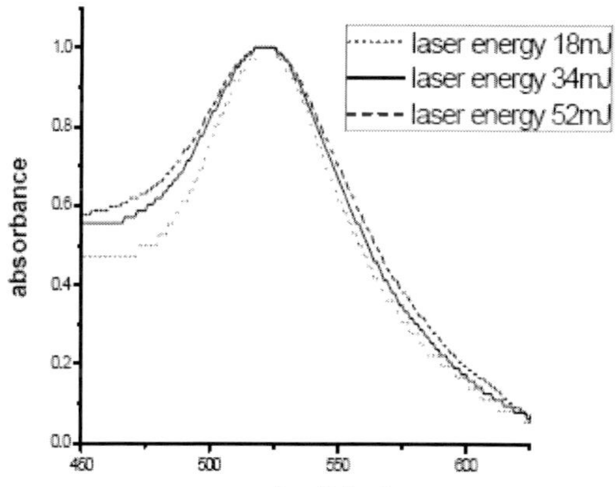

a

Figure 4. (Continued)

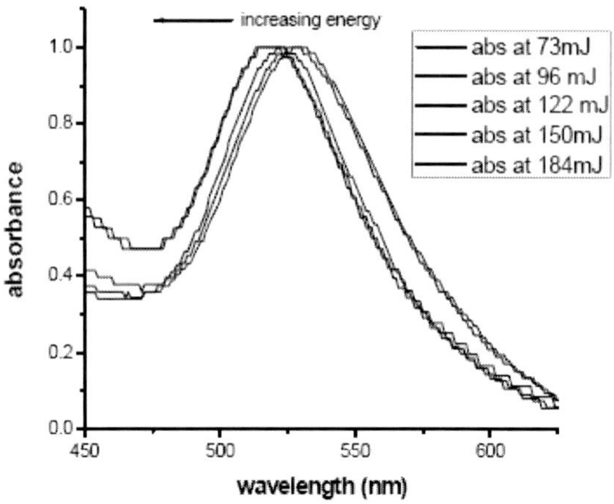

b

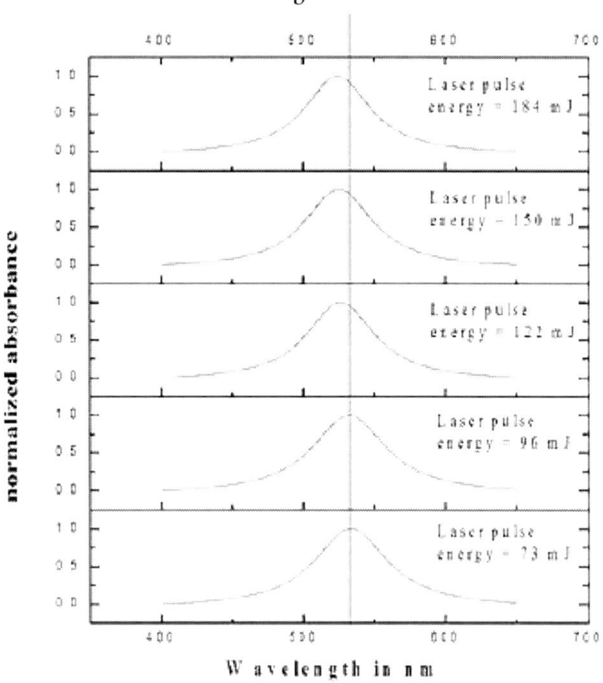

c

Figure 4. Normalized extinction spectra for gold colloids. (a) First 3 colloidal samples generated by laser pulse energies < 70 mJ, (b) Next 5 colloid samples generated by laser pulse energies > 70 mJ. (c) Bell-shaped symmetric part of the extinction spectra for pulse energies > 70 mJ.

It is seen from figure 4 (a) that for laser pulse energy lower than around 70 mJ, the extinction peak remained fairly constant around 519 nm. For the cases when the pulse energy was above 70 mJ, the extinction peak shifted towards the shorter wavelength with increasing laser pulse energy (figures 4 (b), (c) and table 2). Here and also as reported by Doglaev et al. [50] and Tarasenko et al. [12], the curve fits well with the UV-Vis response of spherical metal particles. The SEM images also show that the nanoparticles are spherical (figure 8). According to Mie's theory for spherical particles [45] the Frohlich condition (for resonance) for finite but small metal particles (small enough to retain the contribution upto dipole mode only) was shown to be,

$$\varepsilon' = \mathrm{Re}\,\varepsilon(\omega) = -\left(2 + \frac{12}{5}x^2\right)\varepsilon_m \tag{15}$$

Where $\varepsilon(\omega)$ and ε_m are frequency dependent complex permittivity function of the nanoparticles and the permittivity constant of the medium respectively. $x = (2\pi/\lambda)a$, so as particle size (radius= a) increases λ should increase linearly to satisfy the Frohlich condition for resonance. This is exactly what was obtained in the experiment with a coefficient of correlation r = 0.97 for a linear fit between the extinction peak wavelength and the average particle size (figure 5(a)) for higher laser ablation energy (73 mJ – 184 mJ).

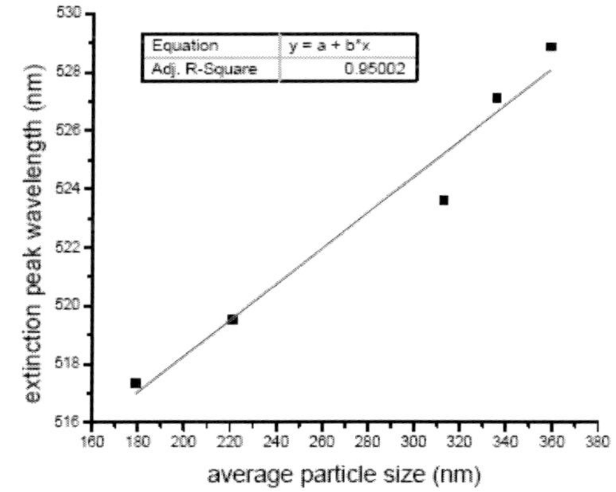

a

Figure 5. (Continued)

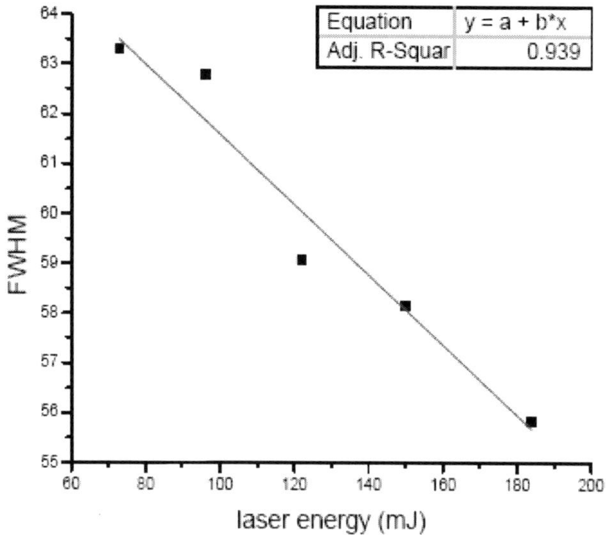

b

Figure 5. (a) Linear correlation between peak wavelength from UV-Vis data and average particle size (diameter) observed by SEM at different laser ablation energy and (b) Full width at half maxima measured from Lorentz fitted curve of the extinction spectra at different laser energy.

From figure 4(a) one can observe that in case of the first three gold colloidal samples (ablated at 18-52 mJ), change of ablation energy is unable to show any shift in the extinction peak which was fixed at 519 nm.

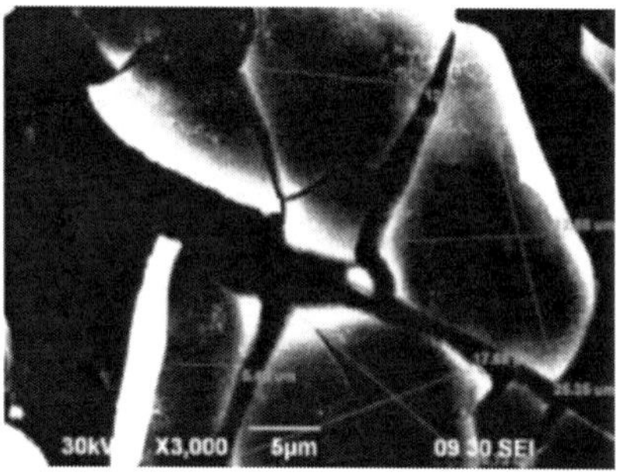

Figure 6. SEM images of fragments observed in nanoparticle generation with laser energies < 70 mJ. Particle sizes are in the micron range as fragments are produced at lower laser ablation energy.

Images taken by scanning electron microscope (figure 6) revealed that the gold particles are in micron range because lower ablation energy formed larger fragments of nanoparticles with occasional smaller nanoparticles. As the particle size increases, scattering gradually becomes dominant over absorption and contributes more to the total extinction efficiency. For such large size particles (micron size) higher order multi-pole mode along with the dipole mode will also start contributing to the extinction of the electromagnetic wave. If one still works with the electrostatic approximation and keeps terms up to the quadrupole mode of surface plasmon for large nanoparticles, then the quadrupole polarizability is given by [44]

$$\mathbf{q} = 4\pi\varepsilon_m a^5 \frac{\varepsilon - \varepsilon_m}{\varepsilon + 3/2\varepsilon_m} \mathbf{E}_o \qquad (16)$$

So the quadrupole resonance condition is

$$\mathrm{Re}[\varepsilon(\omega)] = \varepsilon' = -\frac{3}{2}\varepsilon_m \qquad (17)$$

Therefore, it is evident that although for larger particle, dipole resonance will occur at larger wavelength with associated peak broadening; there will also be resonance peak for scattering due to eq. (17) which is independent of particle size.

Also the absorption peak, though small in intensity for larger particles, occurs near the Frohlich frequency (Frohlich frequency is independent of size) and for gold it occurs at 520 nm [13].

This explains why the extinction peak which is the resultant of absorption and quadrupole scattering peaks, occurred around 520 nm and remained fairly constant for the colloids produced by laser ablation by lower laser fluence.

Information about size distribution of nanoparticles can be obtained from the width of the extinction spectra [46, 47, and 49]. Since the extinction spectra are asymmetric, we have separated the bell-shaped symmetric part from the asymmetric graph. Later we applied Lorentzian fit to the resultant graph for calculating FWHM.

The whole method is graphically described in Fig 7. Figure 7(b) shows that the Lorentzian fit is pretty much the same as the experimentally achieved pattern.

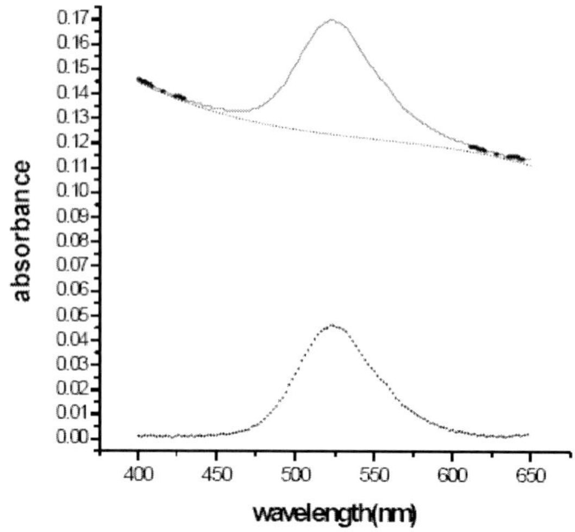

a

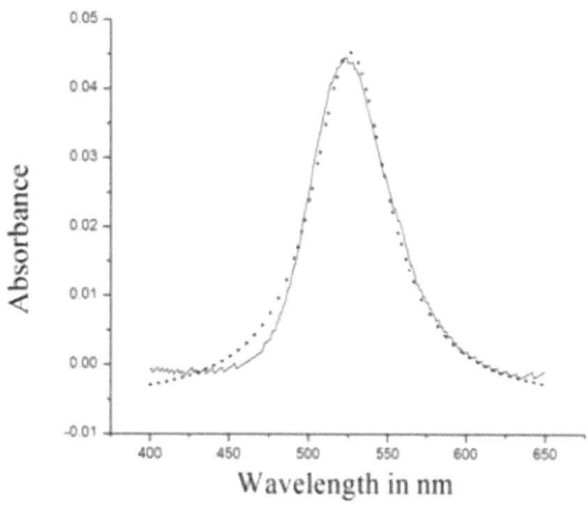

b

Figure 7. An example of measurement of FWHM for extinction spectrum (for laser energy of 122mJ) by applying Lorentz fit: (a) A model of separating the bell shaped symmetric part from the asymmetric extinction spectrum, (b) Lorentz fit (solid curve) to the symmetric curve (dot curve) obtained from figure 7(a).

Scanning electron microscopy (SEM) was used to study the morphology of the colloidal samples ablated at different laser pulse energies. For colloidal sample ablated at a particular energy the microscope was fixed on three different positions at random within the deposited area on the aluminum stub where the gold particles were embedded. For each sample, diameters of all the particles (~ 40 to 60) present in each position on the stub were measured and then the average of all the measurements was recorded. In this way the average diameters of the particles were measured for different ablation energies [19, 30].

Figure 8 shows the SEM images of gold nanoparticles for 5 different ablation pulse energies of the laser. The detail results of our UV-Vis spectroscopy and SEM measurements for colloidal suspension of nanoparticles fabricated at different pulse energies are tabulated in table 2.

SEM Images

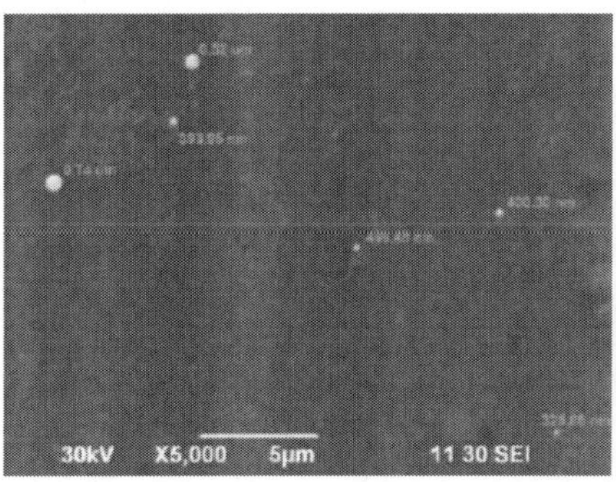

a (i)

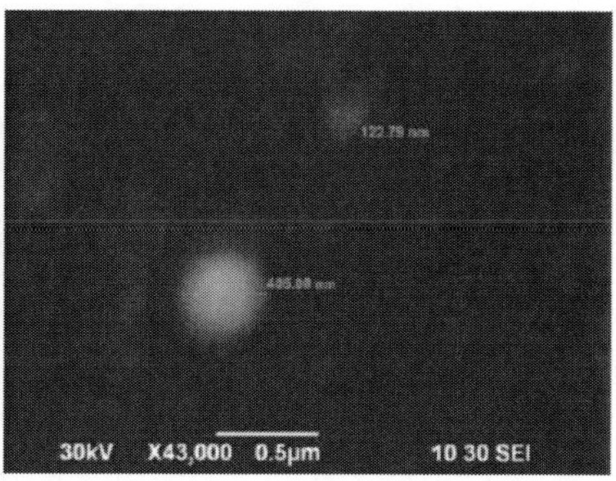

(ii)

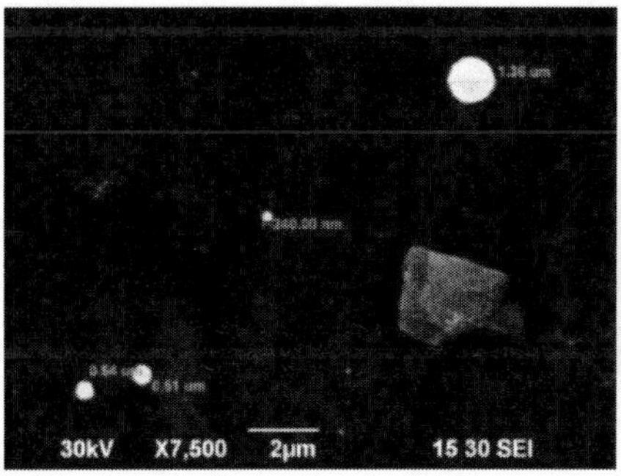

b (i)

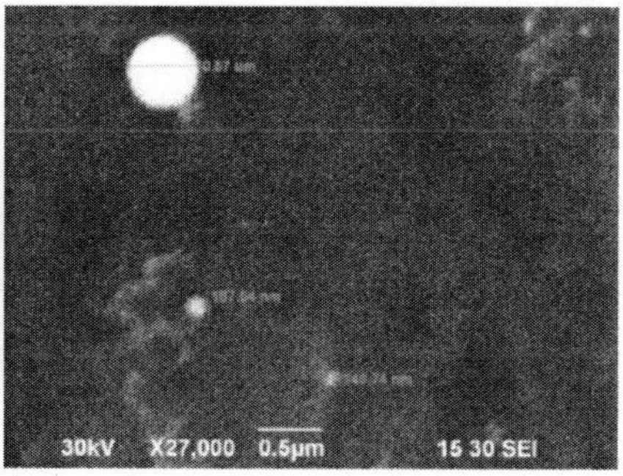

(ii)

Figure 8. (Continued)

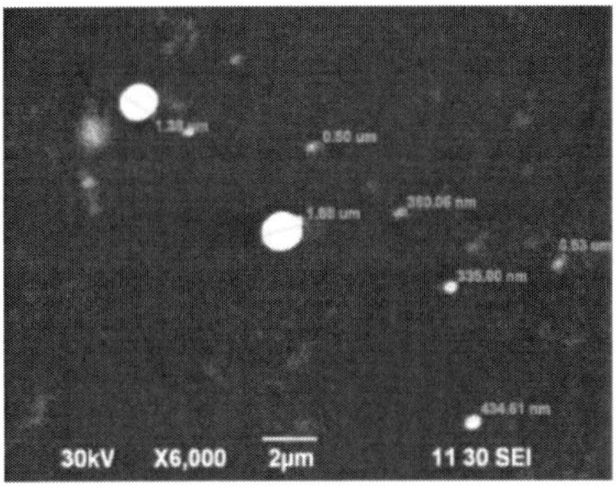

c (i)

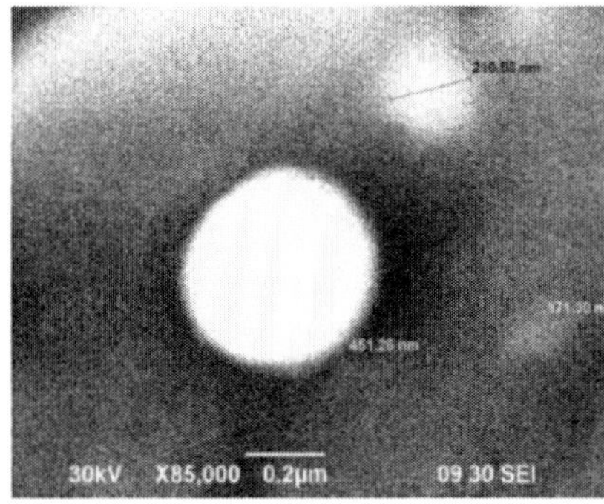

(ii)

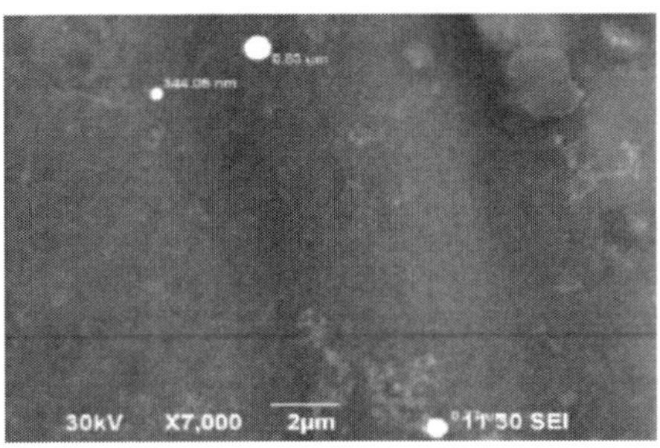

d (i)

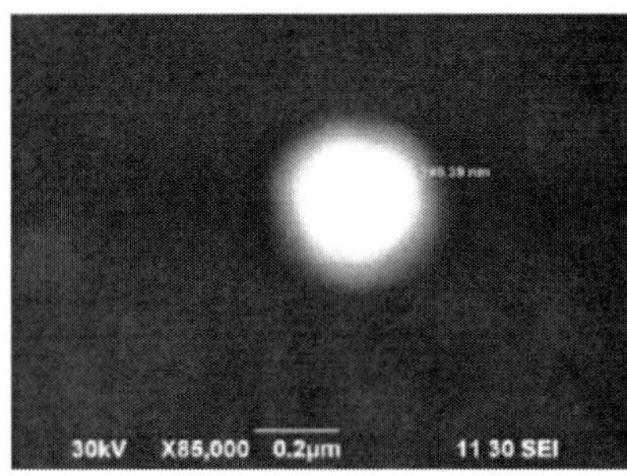

(ii)

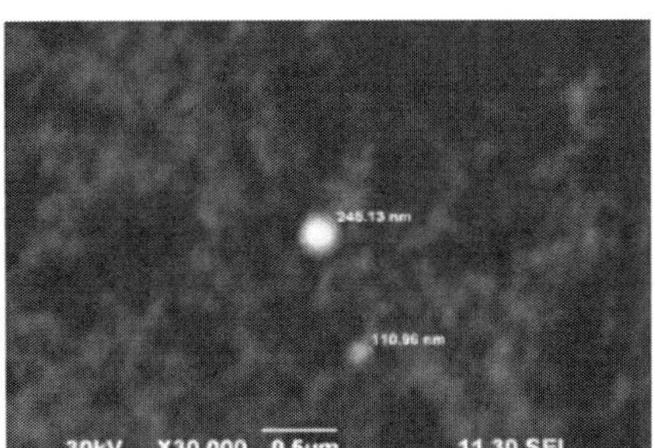

e (i)

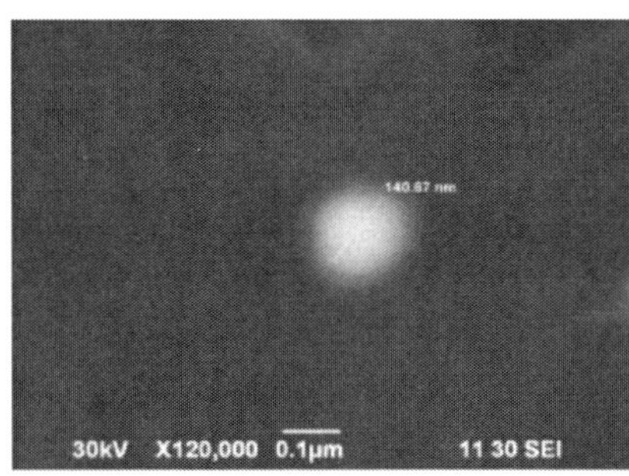

(ii)

Figure 8. For each figure (a)-(e), (i) represents the particles of different sizes and (ii) confirms the spherical shape of particles generated at laser ablation energy of (a) 73mJ, (b) 96mJ, (c) 122mJ, (d) 150mJ, (e) 184mJ respectively.

Table 2. Data for correlation studies between different parameters of gold nanoparticles and ablating laser pulses

Colloidal Sample no.	Laser pulse energy (mJ)	Measurement by UV-VIS spectrometer			Measurement by SEM
		Extinction peak Wavelength (nm)	Extinction peak frequency $*10^{14}$ (Hz)	FWHM of extinction spectrum (nm)	Average particle size (nm)
1	73	528.9	5.67	63.9	359
2	96	527.1	5.69	62.8	336
3	122	523.6	5.73	59.1	313
4	150	519.5	5.77	58.2	221
5	184	517.3	5.80	55.9	179

The extinction spectra obtained for the colloidal solutions are characterized by the peak position which reflects the mean size of the nanoparticles and the FWHM corresponds to the size dispersion [48]. Different parameters of the fabricated nanoparticles were plotted against laser pulse energy, the ablation time etc. to obtain the correlation between them (see table 2).

Figures 9 (a)-(b) show that the extinction peak position and the average nanoparticles size (diameter) both reduce in wavelength and size respectively with the increase of laser pulse energy.

Figure 5 (a) shows a linear correlation between extinction peak position and the average particle size. From figure 5 (b) one can see that when ablation energy increases the particle size dispersion also reduces (as indicated by the decrease of the FWHM of the extinction spectrum). In fact, higher ablation energy leads to larger kinetic energy that increases collision among initially formed large nanoparticles in the condensation process which further reduces the size and makes the size of each particle relatively uniform. The decrease in laser energy results into larger fragments of different sizes and these give rise to large variation of size and hence larger FWHM [19].

We also performed re-ablation on a previously generated gold colloid. This colloid was initially generated at fixed ablation laser pulse energy of 73 mJ for an ablation time of 60 minutes which shows the extinction peak at 520.25 nm. Then it was re-ablated, each time with an increased re-ablation time span of 15 minutes up to a total maximum of 75 minutes. The change in extinction peak position for variation in re-ablation time was recorded [19].

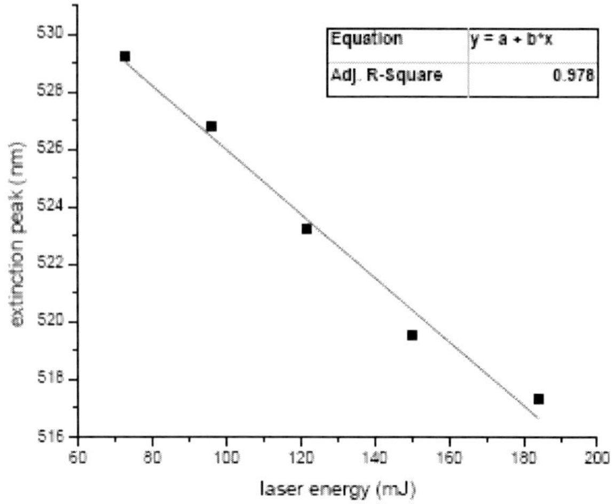

(a)

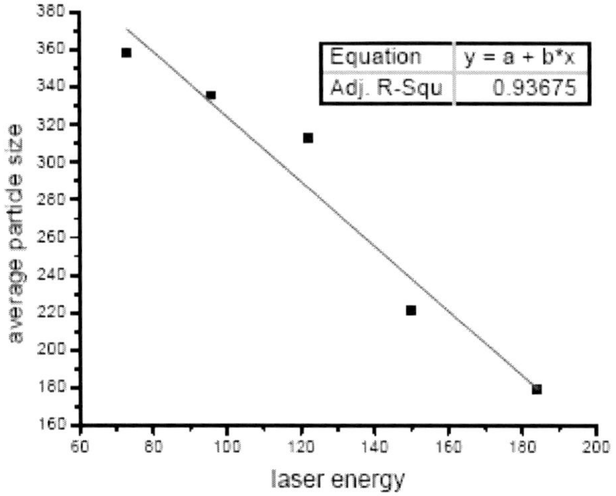

(b)

Figure 9. (a) Extinction peak position of UV spectra and (b) average particle size measured from SEM as a function of laser energy, where the filled squares indicates the experimental data and solid lines indicates linear fit to experimental data. Inset shows the fit equation and parameters.

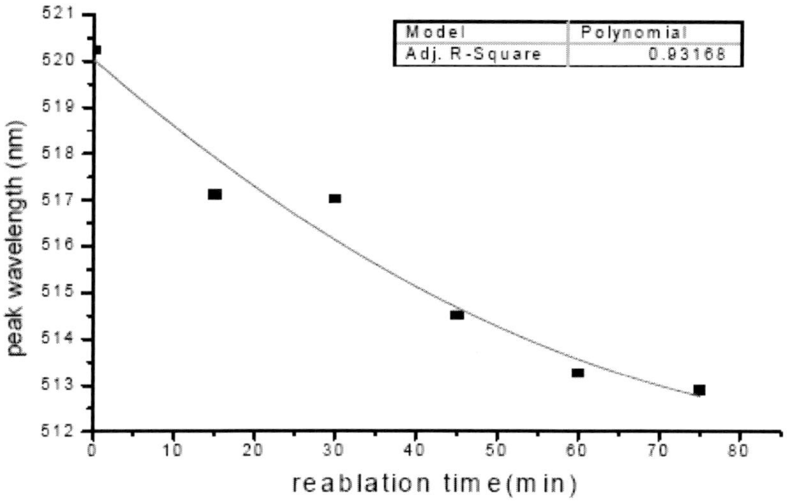

Figure 10. Polynomial fit of peak wavelength with re-ablation time, the ablation laser pulse energy was 73mJ.

Figure 10 shows a clear blue shift of extinction peak position at shortest wavelength of 512.9 nm for the longest re-ablation time. Riabinina et al. [18] observed a blue shift of the absorption spectra of gold colloids with laser irradiation time (with a corresponding decrease in the gold nanoparticle size). They also observed an oxidation of gold upto 26% of its initial concentration formed as a shell on the gold nano-crystal core. They concluded that the blue shift of the absorption peak is most probably due to the superposition of a number of mechanisms, such as size reduction, nano particle surface ionization (injection of electron into the core nanoparticle surface) from the Au-oxide shell induced by laser –matter interaction, absorption and desorption of the surfactant products at the nano particle surface and possibly the decomposition of chemical products leading to the change in the refractive index of the surrounding media. In our work [19] no chemical/surfactant was used. Hence the dominant factor for the blue shift of the extinction peak in the present study is the size reduction of the nanoparticles by re-ablation/fragmentation [19, 49]. The nano particle surface ionization induced by laser-matter interaction may also play some role [19].

We had also carried out the observation of colloid stability with time. Figure 11 shows a comparison between two distinct observations of the extinction spectra of a colloid generated at ablation energy of 96 mJ. The solid line represents the observation when the colloid was freshly prepared and the dashed line for measurement taken after 30 days from preparation. No significant shift was found in the position of the extinction peak but there was a decrease of total area under the curve. The observed stability, also reported by Hahn et al. [48] for gold nanoparticles generated by femtosecond laser, is due to the fact that, as surface to volume ratio increases for nanoparticles, free electrons distribute themselves more uniformly on the surface of the nanoparticles and as a result the surface charge density increases greatly. Now such nanoparticles in the de-ionized water will repel each other and prevent themselves to coagulate. Due to this reason, no significant shift of the extinction peak has been observed as shown in figure 11. No major coagulation was observed in SEM measurement of this sample.

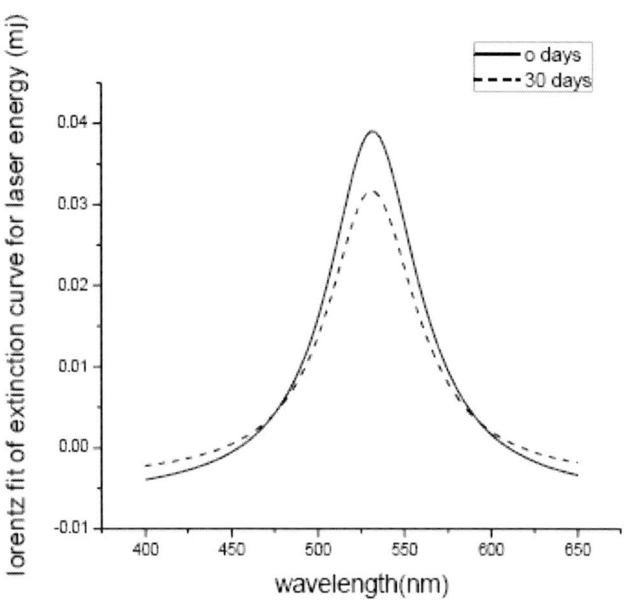

Figure 11. Comparison between two extinction spectra of the same sample observed at a difference of 30 days.

CONCLUSION

In this review paper we have discussed the different techniques of metal nano-particle fabrication with their merits and demerits. As long as the purity of fabricated nanoparticles is concerned, pulsed laser ablation technique is the best. We have discussed the pulsed laser ablation technique, for metal nanoparticle fabrication with particular emphasize on fabrication of gold nanoparticles without the use of any chemical or surfactant.

Good correlations can be obtained between average size of the nanoparticles and laser ablation energy and various other parameters characterizing the nanoparticles. The use of surfactants in water complicates the situation that makes it difficult to study the effect of different parameters of the ablating laser pulse in the fabrication of nanoparticles in colloids [19].

A choice of ablation laser pulse energy and ablation time can lead to significant control in size and the size dispersion of nanoparticles. Higher ablation laser pulse energy leads to more uniform size distribution of nanoparticles (figure 5 (b)). Again a good inverse relation (coefficient of variance = 0.93) has been observed between the absorption peak wavelength and the re-ablation time (figure 10). Possible explanation of the observed blue shift of the absorption peak with re-ablation time is also discussed. Excellent linear correlation, between extinction peak wavelength and average size of the nanoparticles was obtained (with correlation coefficient r = 0.97, figure 5(a)). A good inverse correlation (r = -0.96) was obtained between the average size of the gold nano particles produced and the ablating laser pulse energy (figure 9(b)). An excellent inverse correlation between full width half maxima (FWHM) of the extinction spectra and the laser pulse energy (r = -0.96) was also observed (figure 5(b)).

An explanation was given as to why the extinction peak of gold colloids remained fairly constant at lower laser fluence as observed in the present study and also reported by Simakin et al. [15]. Again the observed blue shift of the extinction peak at higher laser fluence (410 J/cm^2 and above, Table 1) was also explained.

Possible uses of gold nanoparticles, particularly for developing countries were also discussed.

REFERENCES

[1] Eustis S. and El-Sayed . Why gold nanoparticles are more precious than pretty gold:Noble metal surface Plasmon resonance and its enhancement of the radiative and nonradiative properties of nanocrystals of different shapes. *Chemical Soc. Rev*. 35, 209-217, 2006.

[2] Pooja S.S. Fabrication of gold (Au) Nanoparticles in water by laser-induced ablation technique and their characterization and determination

of impurity content in gold ornaments by laser-induced breakdown spectroscopy.[M.S. dissertation]. Dhaka, Bangladesh. Physics department, University of Dhaka, 2007.

[3] Fumitaka M., Kohono J., Takeda Y., Kondow T., *J. Phys. Chem. B* 107, 4218, 2003.

[4] Mafune M.F., Kohono J., Takeda Y., Kondow T., Sawabe H. *J. Phys. Chem. B* 104, 9111, 2000.

[5] Kneipp K., Kneipp H., Monoharan R., Hanlon E.B., Itzkan I., Dasari R.R., Feld M.S. *Appl. Specroscopy*, 52, 12, 1998.

[6] Fumitaka M., Kohono J., Takeda Y., Kondow T., Swaba H. *J. Phys. Chem. B* 105, 5114, 2001.

[7] Russo R.E., Mao X. L., Yoo J. H. and Gonzales J. J. Chapter on '*Laser Ablation' in the book titled* 'Laser Induced Breakdown Spectroscopy' edited by Jagdish P. Singh and Surya N. Thakur, Elsevier.

[8] Kabashin A V, Meunier M. Synthesis of Colloidal Nanoparticles During Femtosecond Laser Ablation of Gold in Water. *Journal of Applied Physics,* 94 ,7941, 2003.

[9] Mafune F, Kohno J-Y, Takeda Y, Kondow T, Sawabe H , Formation of Gold Nanoparticles by Laser Ablation in Aqueous Solution of Surfactant, *Journal of Physical Chemistry, B* 105, 5114, 2001.

[10] Kabashin A V, Meunier M. J. *Photochem Photobiol A. Chem.* 182 330, 2006.

[11] Kazakevich P V, Simakin A V, Shafeev G A, Viau G, Soumare Y, Bozon-Verduraz F. *Appl. Surf. Sci.* 253 7831, 2007.

[12] Tarasenko N.V, Butsen A.V, Nevar E.A, Savastenko N. A. *Appl. Surf. Sci.* 252 4439, 2006.

[13] Scaffardi L B, Pellegri N, Sanctis O de, Tocho JO. Sizing Gold Nanoparticles by Optical Excitation Spectroscopy, *Nanotechnology,* 16 , 158, 2005.

[14] Torchia G.A, Scaffardi L.B, Mendez C, Moreno P, Tocho J.O, Roso L. Optical extinction for determining the size distribution of gold nanoparticles fabricated by ultra-short pulsed laser ablation, *Appl Phys* A 93 967-971, 2008.

[15] Simakin A.V, Voronov VV, Shafeev G.A, Brayner R, Bozon-Verduraz F. *Chem. Phys. Lett.* 348 182, 2001.

[16] Kazakevich P V, Simakin A V, Voronov V V, Shafeev G.A, *Appl. Surf. Sci.* 252 4373, 2006.

[17] Barcikowski S, Hahn A, Kabashin A.V, Chichkov B.N. *App. Phys.* A 87 47, 2007.

[18] Riabinina D, Zhang J, Chaker M, Margot J, Ma D and P. Tijssen. Control of plasmon resonance of gold nanoparticles via Excimer laser irradiation, *Appl Phys, A* 102, 153-160, 2011.

[19] Haider A.F.M.Y., Sengupta S., Abedin K.M., Talukdar A.I. *Appl Phys,* A 105, 487-495, 2011.

[20] Brech F.and Cross C. *Appl. Spectrosc.,* 16, 59, 1962.

[21] Linlor W.I. *Bull. Am. Phys. Soc.,* 7, 440, 1962.

[22] Honig R.E. and Woolson J.R. *Appl. Phys. Lett.,* 2, 138, 1963.

[23] Rosan R.C., Healy M.K. and McNary Jr. W.F. *Science,* 142, 236, 1963.

[24] Berkowitz J. and Chupka W.A. *J. Chem. Phys.,* 40, 2735, 1964.

[25] Zaitsev-Zotov S.V. Martynyuk A.N. and Protasov N.E., *Sov. Phys. Solid State,* 25,100, 1983.

[26] Dietz T.G. Duncan M.A. Powers D.E. and Smalley R.E. *J. Chem. Phys.*74, 6511 1981.

[27] LaiHing K. Wheeler R.G. Wilson W. L. and Duncan M. A. *J. Chem. Phys.* 87, 3401, 1987.

[28] Becker M.E. Broak J.R. Hong Cai. Keto J.W. Lee J. Nicholas W.T. and Glicksman H.D. *NanoStructured Materials,* 10, 853, 1998.

[29] Zeldovich Ya. R. and Raizer Yu. P., Physics of Shock waves and High Temperature Hydrodynamic Phenomena, Dover publications, N.Y., U.S.A., 2007.

[30] Talukdar A I, Sultana P, Haider AFMY, Wahadoszamen M, Abedin KM and Farhad SFM. Power Dependence of Size of Laser Ablated Colloidal Silver Nanoparticles, *European Physical Journal,* D 60 ,295-300, 2010.

[31] Faraday M. Philosophical Transactions of the Royal Society of London 147, 36, 1857.

[32] G. Mie, *Annalen Der Physik* , 25, 377, 1908.

[33] Zsigmondy R. The Chemistry of Colloids, John Wiley & Sons, Inc., New York, 1917. Zsigmondy R. *Colloids and the Ultra-microscope,* John Wiley & Sons, Inc., New York, 1909.

[34] Gans R., *Annalen Der Physik,* 37, 881, 1912.

[35] Zhu Jian., *J. Nanoparticle Res.,* 11, 785-792, 2009.

[36] El-Sayed M.A., *Accounts of Chemical Research,* 34, 257, 2001.

[37] Sharma V. Park K.W. Srinivasarao M. *Materials Science and Engineering Res.* 65, 1, 2009.

[38] Daniel M.C. Astruc D. *Chemical Reviews,* 104, 293, 2004.

[39] Bohren C.F. Huffman D.R. Absorption and Scattering of Light by Small Particles, John Wiley & Sons, Inc., New York, 1983.

[40] Jackson J.D. Classical Electrodynamics, 3rd ed. John Wiley & Sons, Inc., New York, 2001.

[41] Van de Hulst H.C. Light Scattering by Small Particles, Dover Publications, Inc., 1981.

[42] Griffiths D.J., Introduction to Electrodynamics, Prentice-Hall Inc. N.J., U.S.A. 1999.

[43] E. Sadat. "Study of river bed soil of Buriganga, Coral and Beach Sand Samples of Bangladesh by Laser-induced Breakdown Spectroscopy". [M.S. Dissertation]. Dhaka, Bangladesh: Physics department, University of Dhaka, 2006.

[44] Bohren CF and Huffman DR. Absorption and Scattering of Light by Small Particles Wiley-VCH, Weinheim, 1998.

[45] Kelly KL, Coronado E, Zhao LL, Schatz GC. The Optical Properties of Metal Nanoparticles: The Influence of Size, Shape and Dielectric Environment, *Journal of Physical Chemistry B* 107 668, 2003.

[46] Link S and El-Sayed M A. Size and Temperature Dependence of Plasmon Absorption of Colloidal Gold Nanoparticles, *Journal of Physical Chemistry B* 103 4212, 1999.

[47] Katawana A. and Kubo R. Electronic Properties of Fine metallic Particles II: Plasma Resonance Absorption, *Journal of Physical Society of Japan,* 21 1765, 1966.

[48] Hahn A, Barcikowski S, Chichkov BN. Influences on Nanoparticle Production During Pulsed Laser Ablation, *Journal of Laser Micro/Nanoengineering,* 3, 73,2008.

[49] Takami A, Kurita H and Koda S. Laser Induced Size Reduction of Noble Metal Particles, *Journal of Physical Chemistry B* 103 1226,1999.

[50] Doglaev SI, Simakin AV, Voronov V V, Shafeev G A and Bozon-Verduraz F 2002 Nanoparticles Produced by Laser Ablation of Solids in Liquid Environment *Appl. Sur. Sci.* 186 546.

In: Gold Nanoparticles ISBN 978-1-62257-927-3
Editors: A. Jarnagin and L. Halshauser

Chapter 3

SELF ASSEMBLING OF THIOL SURFACTANTS ON GOLD NANOPARTICLES AS NEW TREND

Eid M. S. Azzam*

Applied surfactants laboratory, Petrochemicals department, Egyptian petroleum research institute, Cairo, Egypt

ABSTRACT

Gold nanoparticles have been drawing a lot of attention because of their novel physicochemical properties, especially their special nonlinear optical properties and surface-enhanced Raman scattering effect, and potential applications in biology, catalysis, photocatalysis, and microelectronic.. High affinity of organic functional groups like –SH, -CN, -COOH, $-NH_2$ for metallic surfaces such as Au, Cu, Ag, Pd, Pt, Hg have motivated several research groups to focus on using this phenomena for industrial applications. The self assembling of some synthesized thiol surfactants namely (mercaptopropane-, mercaptohexane-, mercaptooctane- and mercaptodecane sodium sulfonate) on the fabricated gold nanoparticles was investigated. The self assembling of these surfactants on gold nanoparticles was characterized using different techniques. The effect of self-assembling of these surfactants on the size of the gold nanoparticles was studied using TEM images. The growth of the gold nanoparticles was investigated with respect to the increase of alkyl chain in the synthesized thiol surfactants (Azzam and Badawi et al 2009). The fabrication of surface plasmon resonance (SPR) biosensors

* Email: eazzamep@yahoo.com; Fax: 002-02-22747433; Tel: 002-0111-7710207.

using self-assembled monolayers (SAMs) of 1,10-decanedithiol surfactant and adsorbed gold nanoparticles was investigated. Several spectroscopic and microscopic techniques were used to investigate both the SAM and the chemisorption of gold nanoparticles at the SAM surface. Surface plasmon resonance (SPR) was used to study both the adsorption of Bovine serum albumin (BSA) as tested protein to the SAM surface and to the gold nanoparticle-coated SAM (Azzam and Grunwald et al 2009). The self assembling behavior of the synthesized dodecylcysteine hydrochloride surfactant comparing with that of cysteine compound on the prepared gold nanoparticles was confirmed using transmission electron microscope (TEM) measurements. The antitumour activity of the prepared surfactant without and with the gold nanoparticles was investigated (Azzam et al 2008).

1. INTRODUCTION

1.1. Gold Nanoparticles AuNPs

Although gold is the subject of one of the most ancient themes of investigation in science, its renaissance now leads to an exponentially increasing number of publications, especially in the context of emerging nanoscience and nanotechnology with nanoparticles and self-assembled monolayers (SAMs) (Daniel et al. 2004). Whereas the extraction of gold started in the 5th millennium B.C. near Varna (Bulgaria) and reached 10 tons per year in Egypt around 1200-1300 B.C.. When the marvelous statue of Touthankamon was constructed, it is probable that "soluble" gold appeared around the 5th or 4th century B.C. in Egypt and China. In antiquity, materials were used for both pleasant appearance and curative purposes. Colloidal gold was used to make ruby glass and for coloring ceramics and these applications are continuing now. Perhaps the most famous example is the Lycurgus Cup that was manufactured in the 5th to 4th century B.C. It is ruby red in transmitted light and green in reflected light, due to the presence of gold colloids. The reputation of soluble gold until the middle Ages was to disclose fabulous curative powers for various diseases, such as heart and venereal problems, dysentery, epilepsy, and tumors, and for diagnosis of syphilis (Daniel, et al. 2004).

An important class of nanometer-scale materials is colloids and nanocrystals structures that typically are 1-20 nm in diameter and composed of metals, metal oxides, or semiconductors. The small dimensions of these materials give them unique physical properties such as superparamagnetism, fluorescence with high quantum yields (>80%), and depressed useful as stains for analyzing biological samples by electron and optical microscopies, (Jahn 1999)as catalysts for the synthesis of carbon nanotubes and inorganic nanowires, (Gudiksen et al 2000) and as ultrafine magnetic particulates for information storage.

In recent years, colloidal scientists have shifted the focus of their research from micrometer-sized particles to nanometer-sized particles. In particular, studies of inorganic nanostructured materials have received much attention due to their unique optical properties and biocompatible characteristics. Nanoparticles possess characteristics that are significantly different from both their corresponding bulk materials and atomic properties. Many of these properties are related to the size of the nanoparticles; therefore being able to control nanoparticle growth is important (Aliganga et al. 2006), (Liu et al. 2003).

Gold nanoparticles (AuNPs) are the most stable metal nanoparticles, and they present fascinating aspects such as their assembly of multiple types involving materials science, the behavior of the individual particles, size-related electronic, magnetic and optical properties (quantum size effect), and their applications to catalysis and biology. Their promises are in these fields as well as in the bottom-up approach of nanotechnology, and they will be key materials and building block in the 21st century.

The extensive research interest in nanometer dimension inorganic nanoparticles was mainly driven by their broad impact on the emerging disciplines of nanoengineering, nanoelectronics, and nanobioelectronics. Gold nanoparticles, in particular, were the focus of numerous investigations in recent years because of promises offered by their optical, electronic, and chemical properties (Aliganga et al. 2006).

One of the most interesting aspects of metal nanoparticles is that their optical properties depend strongly upon the particle size and shape. Bulk Au looks yellowish in reflected light, but thin Au films look blue in transmission. This characteristic blue color steadily changes to orange, through several tones of purple and red, as the particle size is reduced down to ~ 3 nm (Figure 1). These effects are the result of changes in the so-called surface plasmon resonance, the frequency at which conduction electrons oscillate in response to the alternating electric field of incident electromagnetic radiation. However, only metals with free electrons (essentially Au, Ag, Cu, and the alkali metals) possess plasmon resonances in the visible spectrum, which give rise to such intense colors.

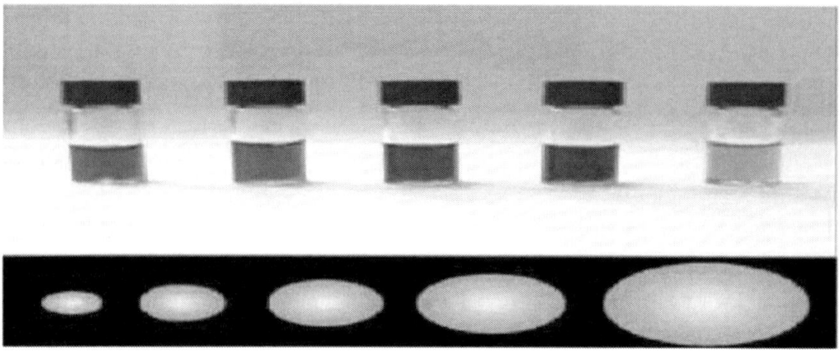

Figure 1. Photographs of aqueous solutions of gold nanoparticles of different sizes.

Elongated nanoparticles (ellipsoids and nanorods) display two distinct plasmon bands related to transverse and longitudinal electron oscillations. The longitudinal oscillation is very sensitive to the aspect ratio of the particles, so that slight deviations from spherical geometry can lead to impressive color changes (Figure 2). Apart from single particle properties, the environment in which the metal particles are dispersed is also of relevance to the optical properties. The refractive index of the surrounding medium, as well as the average distance between neighboring metal nanoparticles also influence the spectral features (Luis 2004). The surface plasmon resonance of gold nanoparticles has broad application and has drawn great attention in recent years.

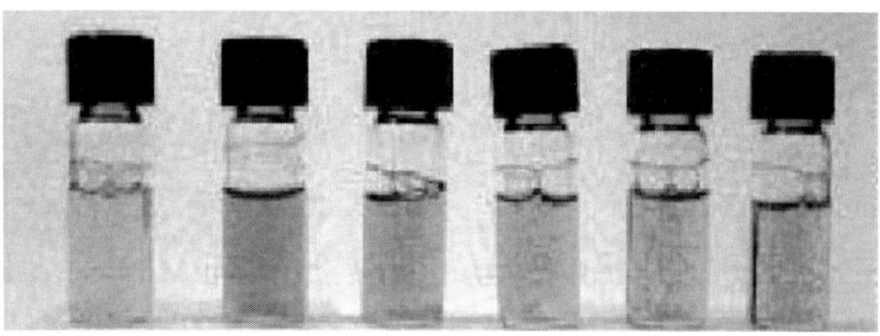

Figure 2. Photograph of aqueous solutions of gold nanorods of various aspect ratios.

Generally, nanoparticles are unstable due to their high surface energy and need to be stabilized against aggregation by suitable surface modifications. Some functional groups, such as cyano (CN), mercapto (SH), and amino (NH_2) are known to have high affinity for gold and hence protective polymers having such functional groups are expected to produce gold nanoparticles with a quit narrow size distribution. Recently, a number of peptides were designed for the stabilization of gold nanoparticles aiming to interact with biomolecules. (Aryal et al 2006) have showed that gold nanoparticles directly conjugated to globular protein, Bovine serum albumin, via S Au bonding provided more surface area for interaction with external species. On the other hand, terminally functionalized alkanethiols and arenethiols have already been demonstrated for their affinity towards gold. Though, the research on the stabilization of gold nanoparticles have grown up, studies on the interaction of gold with biomolecules are still progressing. One aspect of nanoparticles that is different from bulk materials is the percentage of the total number of atoms in the nanoparticle that are interfacial. For example, if gold nanoparticles are assumed to be spherical, a 1.3 nm diameter gold particle has 88% of its atoms on the surface; a 2.0 nm particle has 58% surface atoms;a 5 nm particles has 23% surface atoms; a 10 nm particle has 11.5% surface atoms; a 50 nm particle has 2.3% surface atoms; a 100 nm particle has 1.2% surface atoms; and a 1000 nm particle has 0.2% surface atoms. The electronic states of the interfacial atoms of nanoparticles influence their chemical, electronic, and optical properties. Similarly, the finite size of the cluster can affect its electronic structure.

1.2. The Role of Surfactants in Nanoparticles Formation

More recently, inorganic synthesis has been investigated using gold nanoparticles with ionic liquids such as thiol anionic surfactants (i.e. mercaptoethanesulfonic acid, sodium salt). The ionic liquids have increasingly been drawing attention in research mainly due to their "green solvent" properties. These compounds exist as solventless cations and anions at room temperature. Ionic liquids have a variety of applications due to their tunable properties including polarizability, hydrophobicity, and solvent miscibility which are affected by cation and anion present in the compound. (Azzam and Badawi et al. 2009).

There are two principal strategies for forming metallic and semiconductor nanoparticles: (1) reduction of metal salts (usually in aqueous solutions) and (2) controlled aggregation of zero valent metals (usually in organic solvents). (Murray et al. 2001) To get highly mono-disperse particles, there needs to be rapid nucleation that brings the solution below saturation and then slow, controlled growth until all precursors are consumed. There are two common approaches used to achieve rapid nucleation: decomposition of metal precursors at high temperatures and rapid addition of the reductant. To control the rate of growth and limit aggregation, surfactants are added to the reaction vessel during nanoparticle formation (Puntes et al. 2001). The adsorption of surfactant-like molecules to nucleated nanocrystals lowers the free energy of the surface and, therefore, the reactivity of the particles (Figure 3). The ratio of surfactant to metal precursor can control the size distribution of the nanoparticles. The mechanism by which this ratio controls the nucleation events and limits the growth of the particles is understood in general qualitative terms. The steric bulk of the surfactants provide a physical barrier that prevents the metal surfaces from contacting each other directly. They can also change the surface charge of a cluster and thus change its stability toward aggregation. The combination of the energetic stabilization of the metal surface by the surfactant, the consequences of charge-charge interactions, and the steric repulsion between particles prevents the system from forming aggregates.

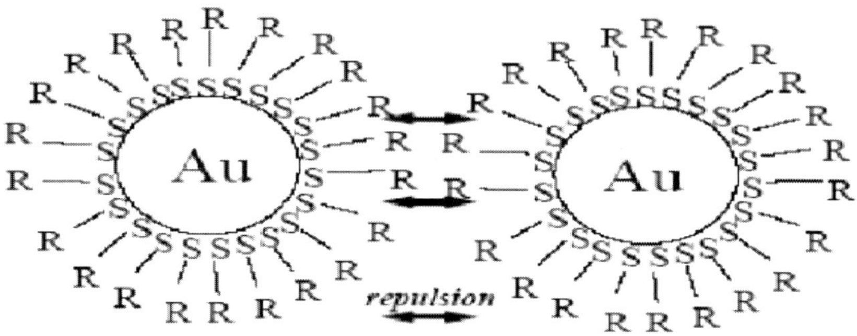

Figure 3. Schematic diagram that illustrate steric bulk of the surfactants provides a physical barrier that prevents the metal surfaces from contacting each other directly.

During the formation of nanoparticles, surfactants interact with the surfaces of the particles in a dynamic equilibrium processs that is, the amount of surfactant on the surface of the particle depends on the relative rates of adsorption and desorption. Surfactants that are chemisorbed to the surface of the particle are less prone to desorption than are physisorbed species. One consequence of a low desorption rate is that particles do not grow rapidly after nucleation. For example, thiols bind via chemisorption to gold nanoparticles and typically limit their sizes to ~ 5 nm (though extremely large particles are also observed in the preparation of small particles) (Jana et al. 2001). In contrast, the majority of surfactants used to stabilize nanoparticles of semiconductors, metal oxides, and some metals (e.g., long-chain acids, amines, phosphines, phosphine oxides, and diols) associate with the surface of particles through van der Waals contacts or weak electronic interactions. These interactions are sufficiently weak that the species readily adsorb and desorb through out the nucleation and growth processes.

The dynamic process allows control over the size and shape of nano-crystals. (Scher et al. 2003) Some classes of surfactants (thiols and dithiols) also can act as reductants in the formation of nanoparticles (Negishi et al. 2003).

1.3. Thiols as a Special Subclass of Surfactants

The chemical reactivity of thiol surfactants make it different from other surfactants. They associate specifically with transition metals to form metal chalcogenides.

Although alkanethiols are used most commonly in the synthesis of gold monolayer protected clusters (MPCs), (Daniel et al. 2004) they are also used in the formation of nanoparticles of many other materials. For gold nanoclusters, the assembly of thiols on their surfaces also can be accompanied by metal etching processes (Prasad et al. 2002). One of the most common routes to gold nanoparticles functionalized with thiols is the Brust-Schiffrin method.

The thiols in this reaction and other related routes are involved in the reduction of gold precursor salts to a Au (I)- thiol polymer (Yee et al. 1999) (Chen et al. 2000).

The Au(I)-thiol intermediates are also useful for forming bimetallic nanoparticles (Au-Pd, Au-Cu, and Au-Ag) via galvanic exchange reactions with thiol-protected metallic nanoparticles. (Song et al. 2003) If dithiols, such as dimercaptosuccinic acid, are used instead of monothiols, Au (III) is reduced completely to Au(0), eliminating the need for any additional reductants in the formation of small (1-2 nm) gold nanoparticles (Negishi et al. 2003).

1.4. Formation of Nanoparticles in the Presence of Thiols

The ω-functionalities of the thiols used to protect nanoparticles determine what solvents (aqueous or organic) can disperse the particles. Some alkanethiols can tolerate the reductive conditions used to prepare nanoparticles and, therefore, can be used to protect the nanoparticles during formation. For example, in the two-phase Brust-Schiffrin method, *n*-alkanethiols and other organic soluble thiols, including a BINOL (1,1´-bi-2- naphthol) derivative, have been used (Choo et al. 2003). Water-soluble nanoparticles are desirable for biological applications, and many preparations have been developed that use thiols with hydrophilic, polar headgroups. For example, mercaptosuccinic acid can serve as a stabilizer during borohydride reduction of HAuCl to give 1-3 nm, water-dispersible gold nanoparticles that are stabilized by the charge-charge repulsion of the carboxylate ions. Glutathione, tiopronin (N-2-mercaptopropionyl-glycine), coenzyme A (CoA), trimethyl (mercaptoundecyl) ammonium, and thiolated derivatives of PEG have all been used as thiol-based water-soluble stabilizers during the formation of gold nanoparticles with a variety of reductants (Zheng et al. 2003).

1.5. Self Assembling of Thiols on AuNPs

It was presented a characterization of self-assembled monolayers of 1-octanethiol (C8-T) and 1,8-octanedithiol (C8-DT)and their binary mixtures immobilized on gold substrates using contact angle investigations, spontaneous desorption time-of-.flight mass spectrometry (SD-ToF-MS)and X-ray photoelectron spectroscopy (XPS). They examined the relationship between the molar ratio of C8-T and C8-DT molecules in solution versus their corresponding molar ratio in the binary mixed SAMs and the ratio between standing up and lying down species of C8-DT. The growth of gold nanoparticles was investigated with respect to the dilution of the nucleation sites in the templating SAM. Influences on the growth velocities were found depending on the amount of available nucleation sites (Aliganga et al. 2006).

(Ding et al. 2006) studied the interactions of thionine with gold nanoparticles by useing UV-vis absorption spectroscopy, fluorescence spectroscopy, transmission electron microscopy, and Fourier transform infrared spectroscopy. With addition of small amounts of gold nanoparticles, the absorption peak intensity of monomers of thionine decreases significantly, whereas that of H-type dimers of thionine increases. With further addition of gold nanoparticles, both the absorption peak intensities of the monomers and dimers of thionine increase, and the increasing magnitude of the peak intensity of the dimers is larger than that of monomers. The addition of gold nanoparticles makes the equilibrium between the monomer and H-type dimer forms of thionine move toward the dimer forms. Due to the binding between thionine and gold nanoparticles, the fluorescence quenching of thionine by gold nanoparticles is enhanced with increasing amounts of gold nanoparticles, and the quenching is both static and dynamic. TEM images show that the addition of thionine results in the formation of gold clusters, and further support the spectral results. Infrared spectroscopic studies show that the nitrogen a toms of both the NH moieties of thionine strongly bind to the gold nanoparticle surfaces through the electrostatic interaction of thionine with gold nanoparticles. For 3 – 6 nm particles, the number of adsorbed thionine molecules per gold particle is about 1.21×10^3. Thionine molecules can not only bind to a particl e to form a compact monolayer via both the NH_2 moieties of thionine, but also bind to two particles via their two NH_2 moieties respectively. (Terasaki et al. 2006) have first prepared a novel bio-nanohybrid electrode consisting of a photosynthetic protein complex, photosystem I (PSI), and gold nanoparticles, and have measured their photocurrent generation properties. The hybrid electrode was fabricated by three steps: sedimentation of gold nanoparticles on a planar gold substrate, followed by modification with 3-mercapto-1-propanesulfonic acid by gold–sulfur self-assembling, and finally modification with PSI by electrostatic adsorption. Even though PSI is one of the proteins and is much larger than usual organic dyes for photocurrent generation, they have succeeded in constructing PSI–gold nanoparticle hybrid electrodes, and in obtaining substantially greater photocurrent responses from PSI as compared with the corresponding planar electrode system. The surfactant dodecylcysteine hydrochloride was synthesised. The surface parameters of the synthesised surfactant were studied using a surface tension technique. The surface parameters show a good surface activity of the prepared surfactant in aqueous solution. The self-assembling behaviour of the synthesized surfactant comparing with that of cysteine compound on the prepared gold nanoparticles was confirmed using transmission electron microscope (TEM) measurements. The effect of self-assembling of this surfactant on the size of gold nanoparticles was studied using TEM images (Azzam, et al. 2008).

1.6. Surface Plasmon Resonance (SPR)

SPR has been demonstrated to be a powerful optical technique for bioaffinity studies at the solid/solution interface. The excitation of surface plasmons by light is denoted as a surface plasmon resonance (SPR) for planar surfaces or localized surface plasmon resonance (LSPR) for nanometer-sized metallic structures. This phenomenon is the basis of many standard tools for measuring adsorption of material onto planar metal (typically gold and silver) surfaces or onto the surface of metal nanoparticles. It is behind many color based biosensor applications and different lab-on-a-chip sensors (van der Merwe 2001). Surface plasmons, also known as surface plasmon polaritons, are surface electromagnetic waves that propagate parallel along a metal/dielectric (or metal/vacuum) interface (Figure 4). Since the wave is on the boundary of the metal and the external medium (air or water for example), these oscillations are very sensitive to any change of this boundary, such as the adsorption of molecules to the metal surface (Garland 1996).

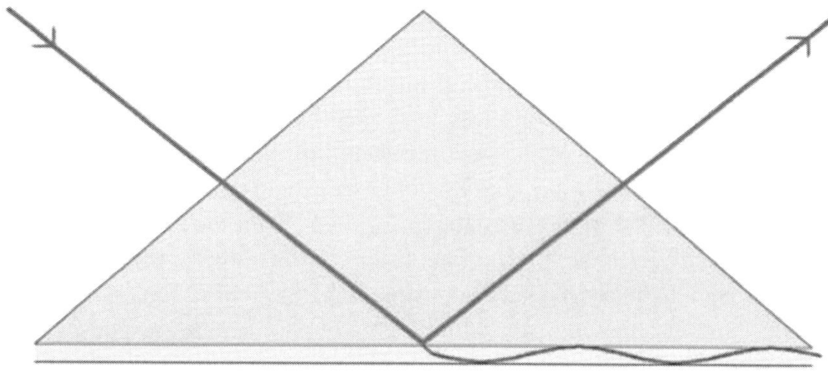

Figure 4. Schematic diagram illustrate the excitation of surface plasmons in a resonant manner.

1.6.1. Applications of SPR

Surface plasmons have been used to enhance the surface sensitivity of several spectroscopic measurements including fluorescence, Raman scattering, and second harmonic generation. However, in their simplest form, SPR reflectivity measurements can be used to detect DNA or proteins by the changes in the local index of refraction upon adsorption of the target molecule to the metal surface. If the surface is patterned with different biopolymers, the technique is called Surface Plasmon Resonance Imaging (SPRI) (Boecker et al. 2007), (Goodrich et al. 2006). For nanoparticles, localized surface plasmon oscillations can give rise to the intense colors of solutions of plasmon resonance nanoparticles and/or very intense scattering. Nanoparticles of noble metals exhibit strong Ultraviolet- Visible absorption bands that are not present in the bulk metal. Shifts in this resonance due to changes in the local index of refraction upon adsorption of biopolymers to the nanoparticles can also be used to detect biopolymers such as DNA or proteins. Related complimentary techniques include plasmon waveguide resonance, QCM and Dual Polarisation Interferometry. Layer by layer self-assembling is one of the first common applications of surface plasmon resonance spectroscopy was the measurement of the thickness of adsorbed self-assembled nanofilms on gold substrates. (Hong et al. 2007) presented a self-constructed immunosensor with angular interrogationbased- surface plasmon resonance (SPR) spectroscopy in order to analyse antigen–antibody interactions. The optimal thickness for the Au film thickness as an active metal for surface plasmons was determined to be approximately 45 nm. They suggested that the sensor can be used as a useful tool for the analysis of immunoreactions.(Huanga et al. 2007) reported that Surface plasmon resonance (SPR) can detect molecules bound to a surface by subtle changes in the SPR angle. By immobilizing probes onto the surface and passing analyte solution through the surface, changes in SPR angle indicate the binding between analyte and probes. Detection of analyte from solution can be achieved easily. By using rolling circle amplification (RCA) and nanogold-modified tags, the signals of analyte binding are greatly amplified, and the sensitivity of this technique is significantly improved. Furthermore, this technique has potentials for ultra-sensitive detection and microarray analysis. This detection technique has great amplification capability. Using 5nm nanogold with 30 min of RCA development time, this proposed protein detection technique shows over 60 times amplification of the original signal. (Yuk et al. 2004) have investigated whether wavelength interrogation-based surface plasmon resonance (SPR) sensors are able to analyze immunoreactions on protein arrays. The results suggested that the wavelength interrogationbased SPR sensors can be used as immunosensors for high- throughput analysis of antigen-antibody interactions on protein arrays.

(Azzam and Grunwald et al 2009) have investigated the fabrication of surface plasmon resonance (SPR) biosensors using self-assembled monolayers (SAMs) and adsorbed gold nanoparticles. The SAM of 1,10-decanedithiol was first fabricated onto a gold substrate. Gold nanoparticles were then chemisorbed onto the SAM surface by bonding with the terminal thiol groups, forming a sensor that can be used to immobilize proteins. Bovine serum albumin (BSA) was used as a test protein in this study. Several spectroscopic and microscopic techniques were used to investigate both the SAM and the chemisorption of gold nanoparticles at the SAM surface. The results are confirmed the covalent bonding of the gold nanoparticles onto the SAM. Surface plasmon resonance (SPR) was used to study both the adsorption of BSA to the SAM surface and to the gold nanoparticle-coated SAM. For SAM surfaces with adsorbed gold nanopartícl es a larger SPR response to BSA than to the sensors wi th a bare SAM is observed.

2. RESULTS AND DISCUSSION

2.1. Synthesis of Gold Nanoparticles

The gold nanoparticles solution (2-4 nm) as shown in Figure 5 was prepared by the reduction of chloroauric acid (HAuCl) with sodium borohydride as mentioned in the previous publication (Azzam and Badawi et al 2009). An 18.5 m l o f water and 0.5 ml of (1 x 10^{-2} mol/ l) trisodium citrate were mixed well, and then 0.5 ml of chloroauric acid (1 x 10^{-3} mol/ l) was added t o the system. The resulting mixture was stirred and then cooled in an ice bath. To the above mixture, 0.5 ml of (0.1 mol= l) sodium borohydride was added slowly and stirred until the colour turned to orange.

Colloidal gold nanoparticles with a diameter of approximately 26 nm as shown in Figure 6 were prepared by reduction of HAuCl using trisodium citrate in aqueous solution (Azzam and Grunwald et al 2009).

2.2. Self Assembling of Some Synthesized Thiol Surfactants on Gold Nanoparticles

The thiol surfactants-coated gold nanoparticles solutions were prepared according to the previous publication (Azzam and Badawi et al 2009). A 20 ml volume of the prepared gold nanoparticles solution was mixed with 5 ml of (1x10^{-5} M) surfactant solution, and then the mixture was stirred for 24 h until the orange color was faded.

The results in figures 7-10 show the self assembling of the prepared mercaptopropane sodium sulfonate (C3), mercaptohexane sodium sulfonate (C6), mercaptooctane sodium sulfonate (C8) and mercaptodecane sodium sulfonate (C10) thiol surfactants on the gold nanoparticles which causes the stabilization of the nano size of these nanoparticles due to the formation of nano shells with these surfactants. It was noticed from figures 7-10 that, the nano size of the gold nanoparticles increases as the alkyl chain moiety of the prepared thiol surfactants increases. This nano size was (1.5-2 nm) with C3, (3-5 nm) with C6, (7-13 nm) with C8 and (8-18 nm) with C10 surfactant as shown in figures 7-10. The TEM images in figures 7-10 showed the large spacing between the gold nanoparticles when they attached to the prepared thiol surfactants. This effect of the prepared thiol surfactants on the nano size of the gold nanoparticles led to the stabilization of the nano size of the gold nanoparticles due to the formation of nano shells between these surfactants and the gold nanoparticles (Azzam and Badawi et al 2009).

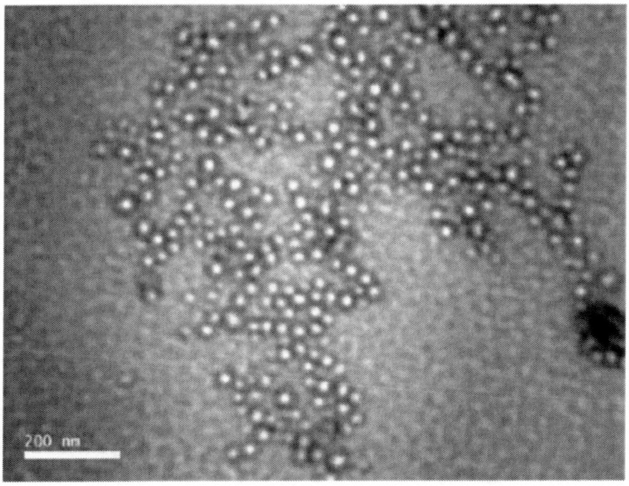

Figure 5. TEM of the gold nanoparticles (2-4nm).

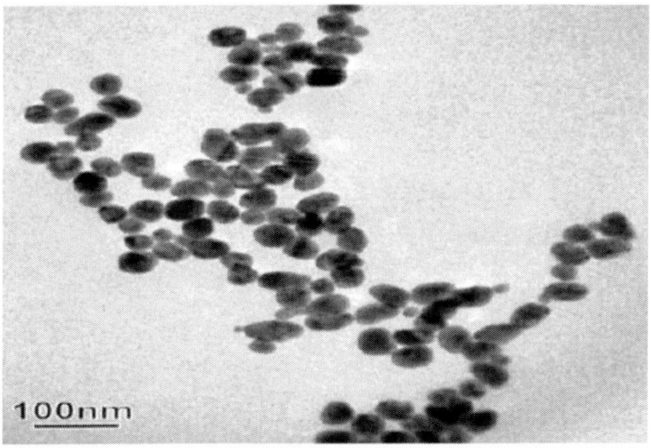

Figure 6. TEM of the gold nanoparticles (26 nm).

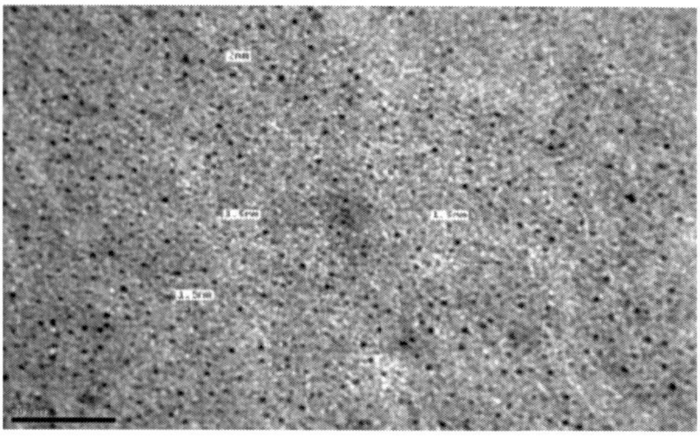

Figure 7. TEM of Mercaptopropane sodium sulfonate(C3).

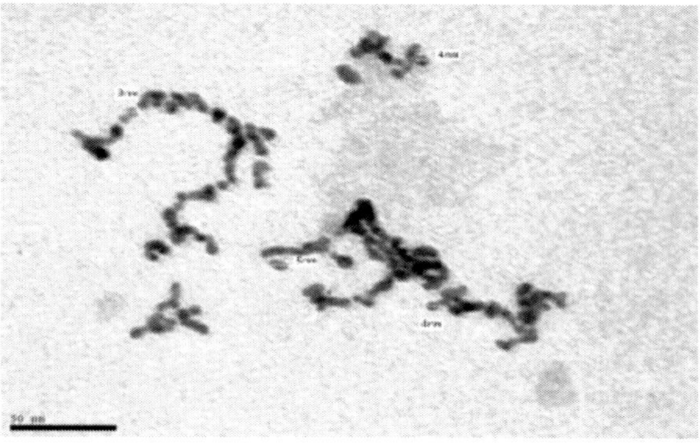

Figure 8. TEM of Mercaptohexane sodium sulfonate(C6).

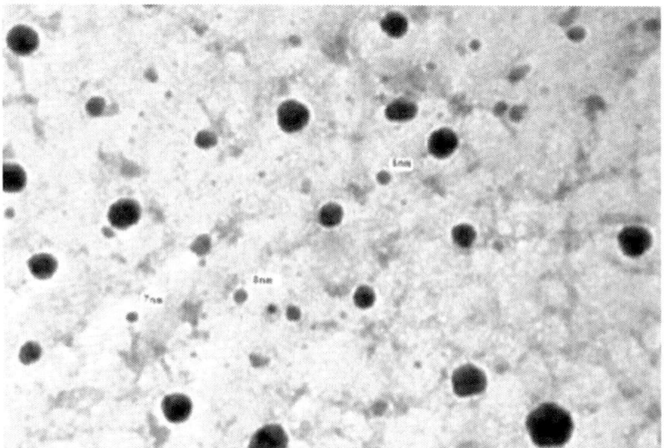

Figure 9. TEM of Mercaptooctane sodium sulfonate(C8).

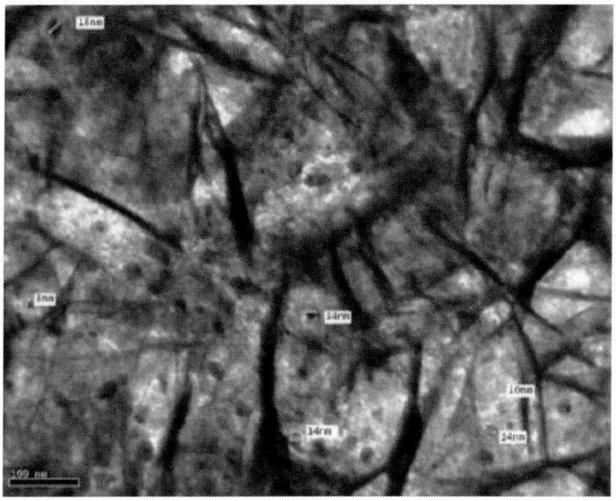

Figure 10. TEM of Mercaptodecane sodium sulfonate(C10).

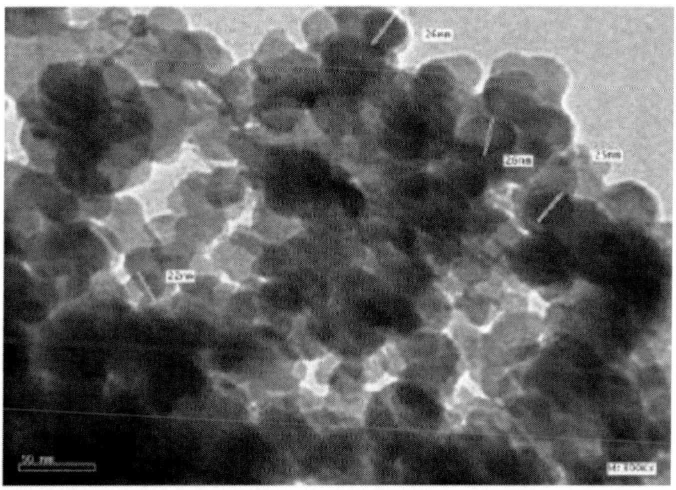

Figure 11. The TEM image of cystein coated gold nanoparticles.

The TEM image in Figure 12 shows the large spacing between the gold nanoparticles when they attached to the dodecylcysteine hydrochloride thiol surfactant molecules which confirmed the self assembling of the prepared surfactant molecules on the gold nanoparticles. The TEM image in Figure 12 also shows the stabilization of the gold nanoparticles size when they attached to the prepared dodecylcysteine hydrochloride surfactant molecules. It was also noticed from Figures 11 and 12 that the gold nanoparticles size is larger when attached to the prepared surfactant molecules (55-111 nm) than with cysteine molecules (22-26 nm) which may be related to the effect of dodecyl alky chain in the prepared surfactant molecule. According to the above TEM results and by comparing between the difference in the size of the individual gold nanoparticles which is (3-6 nm) and the size of these gold nanoparticles with both of cysteine which is (22-26 nm) and with the prepared dodecylcysteine hydrochloride surfactant molecules which is (55-111 nm) these results could be confirm the self assembling of cysteine and dodecylcysteine hydrochloride surfactant molecules on the gold nanoparticles (Azzam et al 2008).

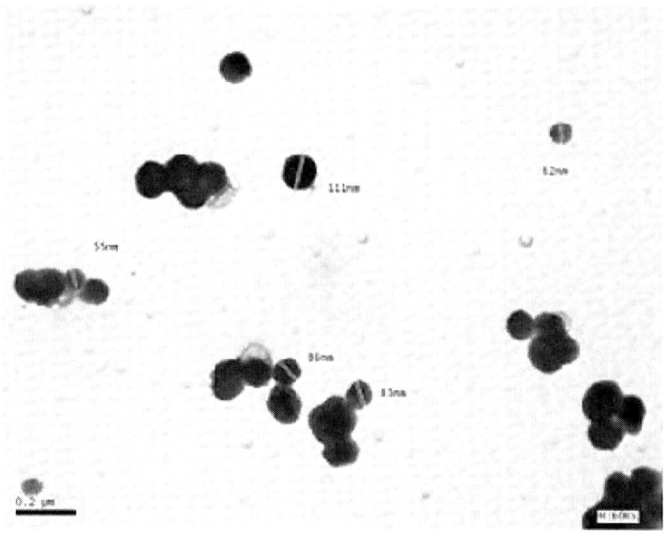

Figure 12. The TEM image of the prepared dodecylcystein hydrochloride surfactant coated gold nanoparticles.

2.3. Fabrication of Biosensors

The fabrication of an SPR biosensor for the detection of the model protein bovine serum albumin (BSA) was illustrated in Scheme 1. Starting from a gold substrate a 1,10-decanedithiol (1,10- DDT) SAM is grown for chemisorption of AuNPs to which the protein is finally adsorbed This biosensor was characterized using infra red reflection – absorption spectroscopy (IRR AS), X- ray photo electron spectroscopy (XPS), atomic force microscopy (AFM), scanning electron microscopy (SEM) and surface plasmon resonance (SPR) as follow (Azzam and Grunwald et al 2009).

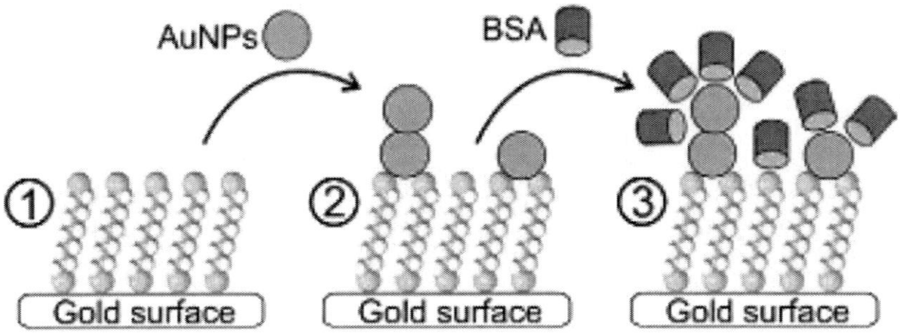

Scheme 1. Schematic illustrating the fabrication process of the SPR biosensor. Step 1: Formation of 1,10-DDT-SAM on gold. Step 2: Chemisorption of AuNP to 1,10-DDT SAM. Step 3: Immobilization of BSA on AuNP.

2.3.1. Chemisorption of the AuNPs onto 1,10-DDT SAM

Figure 13 shows the infrared reflection – absorption spectra of the C – H stretching region of 1,10 - DDT SAMs o n gold with (dotted line) and with out (solid line) AuNPs . For immobilization of the AuNPs the substrate was incubated for 24 h in the AuNP solution. The peak positions of the symmetric (ν sym (CH)) as well as the antisymmetric (ν asym(CH_2)) stretching modes of the methylene groups , appearing at 2851 and 2923 cm^{-1}, respectively, were found to be rather in sensitive to the growth of 1,10 -DDT layers . After attaching AuNPs onto the 1,10 -DDT SAM , the methylene (C H) peak positions showed slight shifts to ν sym = 2846 $cm^{-1}$and ν asym = 2921 cm^{-1}. These small shifts hint at a reorientation of the 1,10-DDT molecules inside the SAM up on AuNP adsorption . Shifts to smaller wave numbers are typically associated with a more crystalline ordering of the molecules in side a SAM. After adsorption of AuNPs to the 1,10 - DDT SAM a reduced absorbance is detected for the stretching modes of the methylene groups which is due to the attenuation of IR-radiation by the adsorbed AuNPs .

XPS was used to estimate the concentration of the free terminal thiol groups of the 1,10-DDT SAM on Au and to investigate the chemisorption of AuNPs onto these groups. XP spectra before and after immersion in t he Au NP solution revealed the chemical composition of the SAM. The binding energy of the major component, Au 4 f at 84.0 eV, was used for calibrating the energy scale of the XP spectra. Survey scans of the 1,10-DDT SAM before and after immersion in the AuNP solution show only signals from carbon, sulfur and gold; no other elements were detected (data not shown). When zooming into the region characteristic for aliphatic carbon the C 1s peak of the 1,10-DDT SAM is resolved at 284.6 eV (Joseph et al 2003).

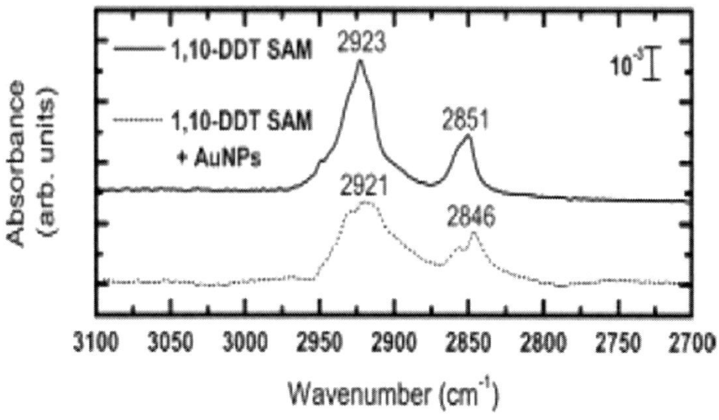

Figure 13. Infrared reflection –absorption spectra of 1,10-DDT SAMswith (dotted) and without (solid) AuNPs.

After immersion in the AuNP solution for 1 h, no shif t in the C 1s peak was observed (data not shown). Figure 14a presents XP spectra zooming into the Au 4f region. From the increase in Au peak intensity af ter incubation in the AuNP solution we conclude that AuNPs are adsorbed to the 1,10-DTT SAM. We further investigated XP spectra zooming into the S 2p region to find out if the AuNPs are chemisorbed (i.e. covalently bound) or physisorbed to the 1,10-DTT SAMs. Two sulfur species of distinct binding energies are present in the S 2p spectrum of the 1,10-DDT SAM shown in Figure 14 b (top spectrum). Before immersion into AuNP solution S 2p peaks at binding energies of ~ 162.3 and ~ 163.4 eV can be discriminated. Based on previous work (Rajalingam et al 2008), these peaks are assigned to Au-thiolate (– S–Au, head group of 1,10-DTT SAM) and to thiol (– S–H, tail group of 1,10-DTT SAM) sulfur species, respectively. After immersion in the AuNP solution, a decrease in the intensity of the SH species (S 2p 3/2 B.E. ~ 163.4 eV) was observed due to the chemisorption of AuNPs to the free terminal SH groups (Figure 14 (b) bottom spectrum). Consequently, due to formation of S –Au bonds the peak characteristic for Au-thiolate species (S2p B.E. ~ 162.5 eV) is increased. Beside the increase of peak intensity this formation of S – Au bonds was accompanied by a slight shift in binding energy of the Au- thiolate peak from S 2p 3/2 B.E. ~ 162.3 eV to S 2p 3/2 B.E. ~ 162.5 eV. The altered sulfur and gold signals are expected due to AuNP adsorption while the unaltered carbon signals indicate that no structural rearrangement of the underlying 1,10-DDT SAM occurs AFM was carried out in order to examine the surface morphology, growth and nucleation of AuNPs at the thiol functionalized 1,10-DDT SA M s . Figure 15 (a– d) presents AFM data of 1,10-DDT SAMs which have been immersed for different incubation times (0, 5 and 25 min and 12 h) into the AuNP solution.

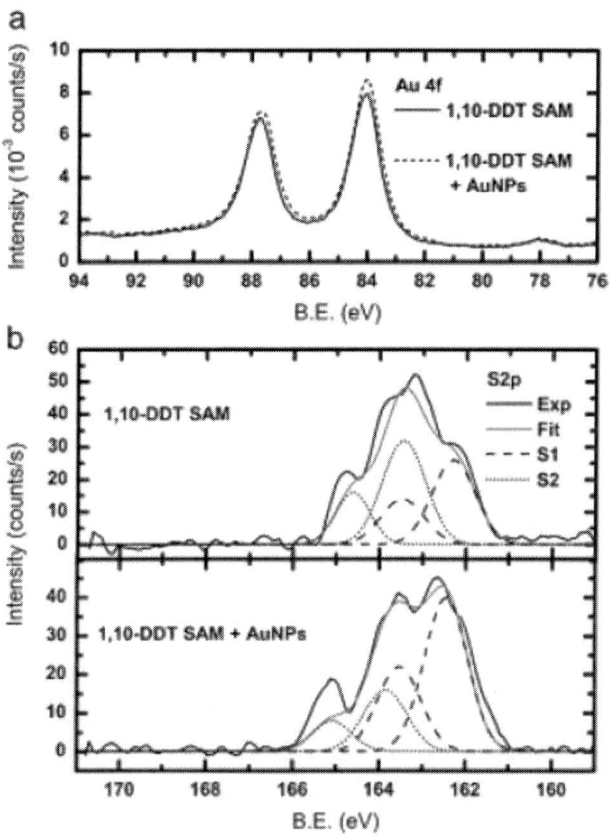

Figure 14. XPS data of the 1,10-DDT SAM on a gold substrate before and after attachment of AuNPs.

Figure 15a shows the surface morphology of a bare 1,10-DDT SAM, revealing small islands with typical heights of 1–2 nm. Figure 15b depicts the SAM surface after 5 min of incubation in the AuNP solution. From this image it seems to be likely that the first chemisorbed AuNPs serve as nucleation sites for cluster growth. Black and white arrows in Figure 15 indicate AuNP aggregates at each immer- sion time and the SAM matrix, respectively. The series of Figure 15 (b–d) demonstrates that the highly clustered AuNP aggregates are formed with prolonged immersion into the AuNP solution. It is clearly recognized that the cluster size increases while the size of the single AuNP is not changed. The tendency towards self aggregation might be explained by the high surface energy of AuNPs. For each sample height profiles are shown along the lines indicated in Figure 15 (a– d). These profiles reveal pronounced differences in the height and width, likely attributable to the process of agglomeration and cluster formation.

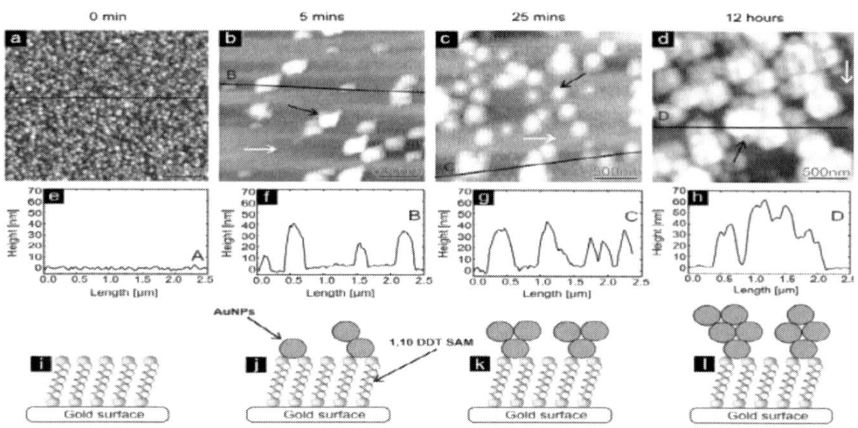

Figure 15. AFM data acquired for (a) the 1,10-DDT SAM and (b – d) the SAM immersed in AuNP solutions for different times. The line scans (e – h) reveal the dependence of height difference for the growth of AuNPs. A schema tic representation of the growth of AuNP island s on the SAMs is depicted in (i–l).

2.3.2. Detection of BSA Protein Using the Fabricated SPR Biosensor

The potential of the AuNP-coated 1,10-DTT SAMs as SPR sensors was used for protein detection in comparison to SAMs without AuNPs coating. Therefore, the model protein BSA adsorbed at different fabrication stages to the sensors surfaces with and without AuNPs. In Figure 16 (a) and (b) the SPR response (shift of the resonance minimum) of each of the two sensor types is plotted against time, while Figure 16 (c) and (d) shows surface plasmon resonance curves. For sensors without AuNPs, two successive injections of BSA (3 min, 0.538 mM ~ 36 g/l) were applied to a naked thiol-terminated 1,10-DDT SAM (see Figure 16 (a) and (c)). A shift in the SPR resonance is detected only during the first injection, indicating that all unspecific binding sites (e.g. hydrophobic patches due to defect sites in the SAM) are blocked by BSA adsorption within the first injection interval. As evidenced by Figure 16 (a) and (c) there is quite a considerable amount of unspecific BSA adsorption to the 1,10-DTT SAM.

The AuNP containing sensors were prepared freshly in situ using the SPR spectrometer: Two successive 3 min lasting injections of AuNPs applied to a 1,10-DDT SAM were already sufficient to ensure AuNP immobilization (see Figure 16 (b) and (d)). Then BSA was adsorbed, as described for the 1,10-DTT SAM surface, to the AuNP-coated sensor surface (3 min BSA injection at a concentration of 0.538 mM ~ 36 g/l).

The amount of adsorbed BSA is evident from Figure 16(b) and (d). When comparing the shift of the minimum in the SPR spectra due to BSA adsorption a more pronounced shift is present with the AuNP based sensors (compare Figure 16(c) and (d)). However, the resonance minimum of the AuNP based sensor became slightly broader which we attribute to the immobilized AuNPs. Nevertheless based on the SPR data the protein detection performance of the AuNP-coated 1,10-DTT SAMs is at least a factor of 2 better than in the case of the naked 1,10- DTT SAMs. We also fabricated sensors using a much longer AuNP adsorption time of 12 h, which leads to large nanoparticle clusters at the SAM surface.

Due to the thicker layer of AuNPs a second strongly shifted and shallower surface plasmon resonance appears (see Figure 17). This second resonance is due to the metallic properties of the thicker layer of AuNPs. Since the shallow resonance is hard to detect no further increase in BSA sensitivity can be achieved in this way. Further experiments to find the optimal performance window for highly sensitive (nanoparticle cluster enhanced) protein detection are on the way.

With the optimal nanoparticle size, density and cluster geometry, the effect of different coatings suppressing unspecific protein adsorption will be studied. Using a longer AuNP adsorption time of 12 h leads to drastic alteration of the surface plasmon resonance curve due to a thicker metallization layer.

The thicker metallization layer does not improve protein detection sensitivity. The nanoparticle surface chemistry used for protein capturing offers an enormous potential for improving the protein detection sensitivity.

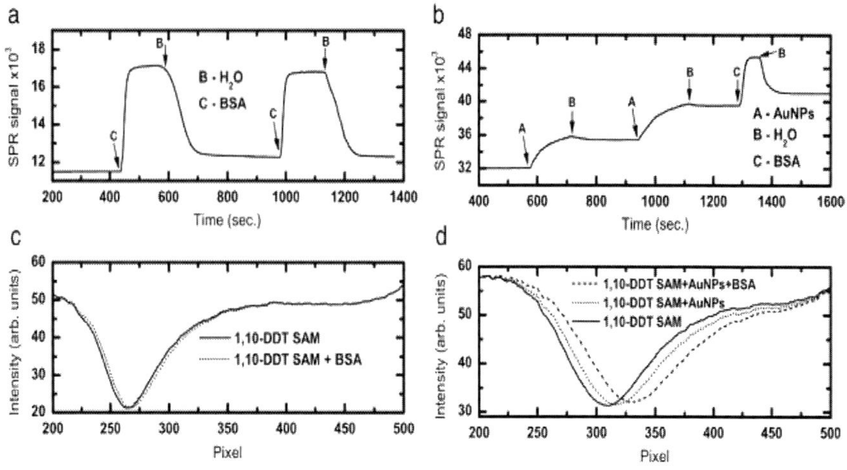

Figure 16. (a, b) SPR sensorgrams and (c, d) SPR spectra of 1,10-DDT SAMs. (a and c) BSA adsorption to 1,10-DDT SAMs. (b and d) Immobilization of AuNPs and subsequent adsorption of BSA yielding a pronounced shift in the SPR signal.

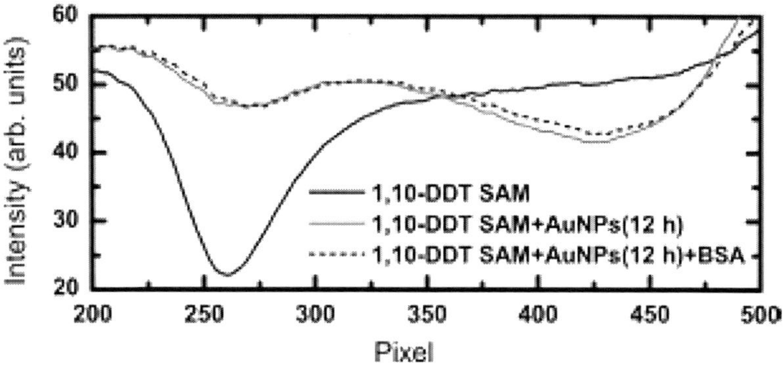

Figure 17. Using a longer AuNP adsorption time of 12 h leads to drastic alteration of the surface plasmon resonance curve due to a thicker metallization layer. The thicker metallization layer does not improve protein detection sensitivity.

2.4. Antitumour Activity

The antitumour activity of the Dodecyl cysteine hydrochloride surfactant without and with the gold nanoparticles was investigated (Azzam et al 2008).

The results show that the antitumour activity of the prepared surfactant was enhanced with the presences of the gold nanoparticles as follow.

2.4.1. Synthesis of Dodecyl Cysteine Hydrochloride – Coated Gold Nanoparticles Solution

A 20 ml of the gold nanoparticles solution was mixed with 5 ml of dodecyl cysteine hydrochloride solution in distilled water. The two mixtures were stirred for 24 h until the orange color had faded.

The TEM image in Fig. 12 shows the large spacing between the gold nanoparticles when they were attached to the dodecyl cysteine hydrochloride surfactant molecules which confirmed the self assembling of the pre- pared surfactant molecules on the gold nanoparticles. The TEM image in Fig. 12 also shows the stabilization of the gold nanoparticles size when they attached to the prepared dodecyl cysteine hydrochloride surfactant molecules.

2.4.2. Antitumour Activity of the Prepared Dodecyl Cysteine Hydrochloride Surfactant-Coated Gold Nanoparticles

The type of cells used is a cell line of the Ehrlich ascites carcinoma (animal breast cancer grows in female albino mice). These cells have been used as an experimental model to evaluate the role of gold nanoparticles in the antitumour activity of the prepared dodecyl cysteine hydrochloride surfactant. The tumouricidal effects of the compound under investigation were tested on the survival of cultured EAC cells, the antitumour activity of the targeted compound was expressed as a percentage of non-viable cells (% NVC). The cytotoxic effects of this compound was dose-dependent, i.e., by increasing the concentration on these compounds in the culture media the percentage of non-viable EAC cells was increased. The antitumour activity dodecyl cysteine hydrochloride surfactant was compared with that of the cysteine compound as shown in Table 1.

From the data in Table 1, it was observed that the cysteine compound show the antitumour activity 10% of non-viable EAC cells at the 25 μg/ml concentration and the maximum cytotoxic effect was 40% at 100 μg/ml concentration, but upon testing the cysteine compound in nano-sized structure with the gold nanoparticles the antitumour activity rose dramatically from 30% of non-viable EAC cells at 25 μg/ml the percentage to 80% of these cells at 100 μg/ml of. Also the antitumour activity of the prepared dodecyl cysteine hydrochloride and dodecylcysteine hydrochloride in nano- sized structure with the gold nanoparticles was tested where the activity against the EAC cells was higher in the case of nano-sized structures. The prepared dodecylcyste- ine hydrochloride surfactant recorded 0% of non-viable EAC cells at 25 μg/ml and 40% of these cells at 100 μg/ml while its nano-sized structure with gold nanoparticles recorded 10% at 25 μg/ml and 50% at 100 μg/ml as shown in Table 1. From the net results in Table 1, it was concluded that the nano-sized structure of the compounds under investigation with the gold nanoparticles was more efficient than the individual compounds towards the antitumour activity by showing a more powerful penetration, spreading and lethal effect with regard to the tested cells. This is related to the very small size of the nanoparticles which allow them to pass through holes in the blood vessels that surround a tumour (Russell 2003).

Table 1. The Antitumour activity of cysteine compound, the prepared dodecylcysteine hydrocloride surfactant and their nano sized structure with the colloid gold nanoparticles (AuNPs)

Sample	% inhibition of cell viability		
	25 μg/ml	50 μg/ml	100 μg/ml
Cysteine	10 %	20 %	40 %
Cysteine with AuNPs	30 %	50 %	80 %
dodecylcysteine hydrocloride	0 %	10 %	40 %
dodecylcysteine hydrocloride with AuNPs	10 %	20 %	50 %

References

Aliganga, A. A.; Duwez, A. S.; and Mittler, S.; *Organic Electronics* 7, 337 (2006).

Aliganga, A. A.; Duwez, A. S.; and Mittler, S.; *Organic Electronics* 7, 337 (2006).

Aryal, S.; Remant, B.K.C.; Dharmaraj, N.; Bhattarai, N.; Kim, C. H.; and Kim, H. Y.; *Spectrochimica Acta Part A* 63, 160 (2006).

Azzam, E. M. S. and Morsy, S. M. I.; *J. Surfact. Deterg.* 11, 195 (2008).

Azzam, E. M. S., Badawi, A. M. and Alawady, A. R. E. *JDST,* 30 (4), 540 (2009).

Azzam, E. M. S., Grunwald, Ch. , Bashir, A., Shekhah, O. , Alawady, A. R. E., Birkner, A. and Wöll, Ch., Thin Solid Films, 518 (1), 387-391 (2009).

Boecker, D.; Zybin, A.; Horvatic, V.; Grunwald, C.; and Niemax, K.; *Anal. Chem.,* 79,702 (2007).

Chen, S. W.; Templeton, A. C.; and Murray, R. W.; *Langmuir* 16, 3543 (2000).

Choo, H.; Cutler, E.; and Shon, Y. S.; *Langmuir* 19, 8555 (2003).

Daniel, M. C.; and Astruc, D.; *Chem. Rev.* 104, 293 (2004).

Ding, Y.; Zhang, X.; Liu, X.; and Guo, R.; *Colloids and Surfaces A: Physicochemical and Engineering Aspects,* 290, 82 (2006).

Garland, P.; Q. *Rev. Biophys.* 29, 91 (1996).

Goodrich, T. T.; Wark, A. W.; Corn, R. M.; and Lee, H. J.; *Methods in Molecular Biology* 328, 113 (2006).

Gudiksen, M. S.; and Lieber, C. M.; *J. Am. Chem. Soc.* 122, 8801 (2000).

Hong, D.G.; Kim, T.W.; Kim, K. B.; Yuk, J. S.; and Ha., K. S.; *Meas. Sci. Technol.* 18, 1367 (2007).

Huang, X.; El-Sayed, I. H.; Qian, W.; and El-Sayed, M. A.; *Nano Letters* 7(6), 1591 (2007).

Jahn, W. J.; *Struct. Biol.* 127, 106 (1999).

Jana, N. R.; Gearheart, L.; and Murphy, C. J.; *Adv. Mater.* 13, 1389 (2001).

Joseph, Y., Besnard, I., Rosenberger, M., Guse, B. Nothofer, H.-G., Wessels, J.M., Wild, Knop-Gericke U. A ., Su, D., Schlogl, R., Yasuda, A ., Vossmeyer, T., *J. Phys. Chem. B* 107 7406 (2003).

Liu, S. Q.; Ju, H. X.; *Anal. Lett.* 36 (1), 1 (2003).

Luis, M. L.; *Materials Today* 7 (2), 26-31 (2004).

Murray, C. B.; Sun, S.; Gaschler, W.; Doyle, H.; Betley, T. A.; and Kagan, C. R.; *IBM J. Res. Dev.* 45, 47 (2001).

Negishi, Y.; and Tsukuda, T.; *J. Am. Chem. Soc.* 125, 4046 (2003).

Prasad, B. L. V.; Stoeva, S. I.; Sorensen, C. M.; and Klabunde, K.; *Langmuir* 18, 7515 (2002).

Puntes, V. F.; Krishnan, K. M.; and Alivisatos, A. P.; *Science* (Washington, D.C.) 291, 2115 (2001).

Rajalingam, K., Strunskus, T., Terfort, A., Fischer, R. A., Woll, C., *Langmuir.* 24 7986 2008).

Russell, D. R., *Pharm Technol,* 17 (2003).

Scher, C.; Manna, L.; and Alivisatos, A. P.; *Philos. Trans. R. Soc.* London, *Ser. A: Math. Phys. Eng. Sci.* 361, 241 (2003).

Song, Y.; Huang, T.; and Murray, R. W.; *J. Am. Chem. Soc.* 125, 11694 (2003).

Terasaki, N.; Yamamoto, N.; Hiraga, T.; Sato, I.; Inoue, Y.; and Yamada, S.; *Thin Solid Films* 499, 153 (2006).

Van der Merwe, P.; Practical Approach series, *Oxford University Press,* 137 (2001).

Yee, C. K.; Jordan, R.; Ulman, A.; White, H.; King, A.; Rafailovich, M.; and Sokolov, J.; *Langmuir* 15, 3486 (1999).

Yuk, J.; and Ha, K.; *Journal of the Korean Physical Society* 45 (4), 1104 (2004).

Zheng, M.; Davidson, F.; and Huang, X.; *J. Am. Chem. Soc.* 125, 7790 (2003).

In: Gold Nanoparticles ISBN 978-1-62257-927-3
Editors: A. Jarnagin and L. Halshauser

Chapter 4

FORMATION OF GOLD NANOPARTICLES INSIDE THE CORONA OF AMPHIPHILIC TRIBLOCK COPOLYMER MICELLES

Anastasia Meristoudi and Stergios Pispas
Theoretical and Physical Chemistry Institute,
National Hellenic Research Foundation, Athens, Greece

ABSTRACT

Gold nanoparticles (Au NPs) have attracted significant attention due to their unique optical properties which differ from those observed in the bulk material. Their size, shape and morphology, as well as the surrounding medium are the key factors that enhance and tune these properties, allowing for the application of AuNPs in photonic devises, biosensors, nanoreactors and in biomedicinal applications. Usually, amphiphilic block copolymer micelles are used for the formation of AuNPs, either inside their cores or into their corona, depending on the envisioned application. In this work, poly[(methacrylic acid)-b-(styrene)-b-(methacrylic acid)] (PMAA-PS-PMAA) triblock copolymer was utilized for the *in situ* synthesis of Au NPs inside the corona of the copolymer micelles in aqueous media. Stoichiometric equivalents of $HAuCl_4$ were added as the metal precursor, which were preferentially dissolved and coordinated to the coronas of the block copolymer micelles and were subsequently reduced to Au NPs without addition of any reducing agent. Therefore, the block copolymer acts as the nanotemplate and the carboxylate groups as the reducing agents. UV-vis absorption spectroscopy was employed in order to follow the Au NPs formation kinetics. Dynamic light scattering (DLS) was utilized for studying the colloidal stability and structural response of PMAA-PS-PMAA/Au NPs hybrid nanostructures, as a function of the pH, the ionic strength and the temperature of the solution. Additional information on the structure of the nanocolloidal system, the polymer-inorganic material interactions and metal nanoparticles characteristics were obtained by the use of atomic force microscopy (AFM) and transmission electron microscopy (TEM).

Keywords: Metal nanoparticles, polymer induced reduction, gold nanoparticles formation kinetics, triblock copolymer micelles

INTRODUCTION

Metal nanoparticles, and more specifically gold nanoparticles, have attracted significant attention during the last decades because they exhibit novel optical, electrical, catalytic and photonic properties which are strongly dependent on their size, shape and morphology, as well as on the surrounding medium [1,2,3]. Typically, metal nanoparticles are synthesized by incorporation into polymeric matrices, such as copolymers or dendrimers, followed by reduction of the metal ion [4,5]. Metal nanoparticles can be in situ synthesized either inside the dense core of polymeric micelles or around the corona of the micelles allowing thus for exploitation of all the AuNPs' unique properties. It has been observed that the nanoscale domains formed by the copolymers provide the ideal environment for nanoparticle growth. Additionally, the copolymer architecture influences the nanoparticle surface modification and the positioning of the nanoparticles within the hybrid nanostructure [6].

In recent years researchers have been challenged to utilize non-toxic chemicals, environmental friendly solvents and renewable materials towards the synthesis of gold nanoparticles with controlled size and shape [7]. Amino and hydroxyl containing polymers have been utilized towards this synthetic goal. AuNPs formation has been observed through reduction of Au^{3+} to Au^0 due to electrons transferred from the amine to the metal ion for the case of polymers containing amino groups [8]. Main chain ether polymers containing a hydroxyl terminal functional group form pseudocrown ethers in the presence of metal ions resulting in encapsulating them and facilitating their reduction to metallic nanoparticles [9].

Alexandridis et al. showed that poly[(ethylene oxide)-b-(propylene oxide)] (PEO-PPO) block copolymer acts as a very efficient reductant and colloidal stabilizer in the single-step synthesis and stabilization of gold nanoparticles from $HAuCl_4$ in aqueous solution, at ambient temperature and in the absence of any reducing agent or energy input [10]. In another study they proved that PEO-b-PPO could act as the reducing, morphogenic and stabilizing agent in the synthesis of Ag nanoparticles [11]. Going a step further, and using a mixture of triblock copolymers, as a method to adjust the hydrophobic/hydrophilic environment in order to tune the size and shape of AuNPs [12], they used the triblock copolymer PEO-PPO-PEO for the synthesis of gold nanoparticles of the desired size and shape [13]. The shape and the size of the gold nanoparticles could be defined by the block copolymer molecular weight, the PEO or the PPO length, the PEO/PPO block ratio and the concentration of the metal precursor [14].

Khullar et al [15] also used PEO-PPO-PEO triblock copolymer for the in situ synthesis of AuNPs in the absence of any reducing agent. Two different micellar triblock copolymers were studied for reducing $HAuCl_4$, the PEO_{76}-PPO_{26}-PEO_{76} and PEO_{17}-PPO_{60}-PEO_{17}, commercially known as F68 and P103 respectively. They proposed that the mechanism of Au^{3+} reduction involves sequential redox reactions on the surface of the PEO ether backbone. They also suggested that the particle size could be influenced by the solvent, the temperature and the PEO/PPO ratio, while the shape could be controlled by the $HAuCl_4$ concentration. Fixing the metal nanoparticles inside the dense cores of polymer micelles often hinders some of their very important properties, such as catalytic or sensing ones. Various studies have focused on the synthesis of metal nanoparticles on the corona or the shell of polymer micelles. Shi was the first one who proposed the synthesis of a hybrid material consisting of a poly(ethylene oxide) corona, a poly(vinyll pyridine)/AuNPs inner shell and a poly(styrene) core [16]. First, hybrid micelles of the block copolymer PS-b-P4VP were synthesized having Au nanoparticles fixed on the P4VP corona. Then poly[(ethylene oxide)-b-(vinylpyridine)] block copolymer was added resulting in the formation of a multifunctional hybrid nanostructure consisting of a PS core, a P4VP/Au pH sensitive inner shell and a PEO corona. However, in this case the use of a reducing agent for the formation of AuNPs was necessary. Along these lines, AuNPs have been synthesized in the inner shell of poly[(ethylene glycol)-b-(styrene)-b-(vinylpyridine)] triblock copolymer [17]. The hybrid nanostructure is pH sensitive and more specifically under alkaline conditions channels are formed through the hybrid P4VP/Au/PEG inner shell allowing interaction of the dense hydrophobic PS core with the environment, enhancing thus the catalytic properties of AuNPs. Other examples involve the use of the pH sensitive and thermosensitive poly(isopropylacrylamide) PNIPAM [18,19] as the block of a block copolymer directly involved for AuNPs synthesis. Depending on the experimental conditions, i.e. basic or acidic pH or temperature, AuNPs can be synthesized around the corona of well defined micelles. Another synthetic strategy for the formation of AuNPs on the corona of micelles involves the micellar structural inversion due to changing the solvent from micelles already containing AuNPs in their core [20,21]. Nevertheless, in all these studies organic solvents and severe reducing agents are necessary for the formation of AuNPs. However, only a few studies exist, to the best of our knowledge, that the metal nanoparticles are in situ synthesized at the periphery of the polymer micelles, so their properties are fully exploited, using only ambient and environmentally friendly conditions. Recently, our group utilized poly[(sodium sulfamate/carboxylate-isoprene)-b-tert-butylstyrene] (CSI-BSI) diblock copolymer for the in situ polymer mediated synthesis of AuNPs at the corona of CSI-BSI micelles [22]. $AuCl_4^-$ was dissolved and preferentially coordinated at the corona of each micelle and subsequently reduced to Au^0 by the amino groups of the corona block. The CSI-BSI/Au NPs hybrid nanocolloid was complexed with lysozyme resulting in a multifunctional protein carrier nanosystem. In this study we utilized poly[(methacrylic acid)-b-(styrene)-b-(methacrylic acid)] (PMAA-b-PS-b-PMAA) triblock copolymer micelles for the synthesis of Au nanoparticles in water. Since water is a selective solvent for the PMAA block, it leads to the formation of micelles with a hydrophobic PS core and a hydrophilic PMAA corona. $HAuCl_4$ was used as the metal precursor and was preferentially dissolved and coordinated to the corona of the micelles. The reduction of Au^{3+} to Au^0 was facilitated by the block copolymer, without the use of any reducing agent. Various experimental conditions such as pH, ionic strength, temperature, metal precursor concentration were carefully considered in order to examine their effect on the stability and responsiveness of the PMAA-PS-PMAA/Au NPs hybrid nanoclusters.

EXPERIMENTAL

Materials

Au 99.99% was dissolved in HCl/HNO_3 (3:1) leading to a red residue, which was then diluted into ethanol in order to prepare a 0.03M solution of

$HAuCl_4$. Poly[(methacrylic acid)-b-(styrene)-b-(methacrylic acid)] (PMAA-PS-PMAA) triblock copolymer (Figure 1) was synthesized by anionic polymerization and post polymerization functionalization hydrolysis of the precursor poly[(tert-butylmethacrylate)-b-(styrene)-b-(tert-butylmethacrylate)] (PtBMA-PS-PtBMA) triblock copolymer (molecular characteristics of the final triblock copolymer: 12wt% PS, $M_w = 4.6x10^4$, $M_w/M_n = 1.03$).

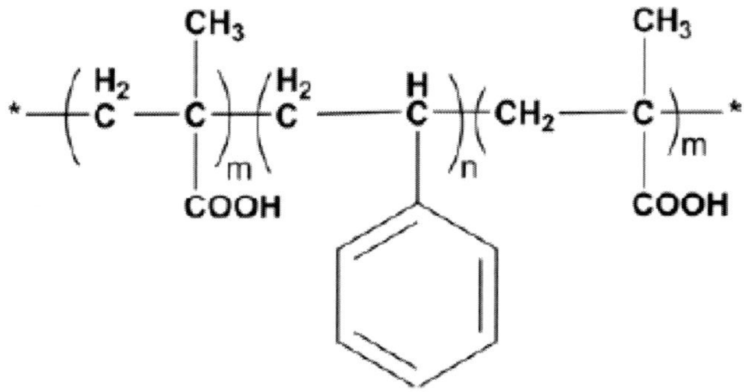

Figure 1. Chemical structure of PMAA-PS-PMAA triblock copolymer used in this study.

Micelle Formation

The triblock copolymer PMAA-PS-PMAA was dissolved in distilled water (pH~6.8) in order to prepared polymer solutions of 10 mg/mL concentration. Further dilution of the stock solution led to the formation of solutions of lower polymer concentration, ranging from 1 to 10 mg/mL. All solution preparations took place at room temperature. Water is a bad solvent for PS, leading to the formation of micellar aggregates with a PS core and a PMAA corona.

Hybrid Metal-Polymer Solutions

Hybrid micelles of the triblock PMAA-PS-PMAA copolymer were obtained by addition of 0.5 and 0.25 equiv. of $HAuCl_4$ per acid unit of the copolymer. Au was introduced into the micelles' coronas due to coordination with the MAA units and subsequent reduction of Au^{3+} to metallic Au occurred, leading to purple solutions. A schematic illustrating the preparation protocol is given in Figure 2. The responsiveness of the PMAA-PS-PMAA/Au NPs hybrid colloid towards ionic strength changes was evaluated by addition of various quantities of NaCl (in the range of 0.03M to 0.2M). In order to investigate the pH effects on the structure and stability of already formed Au nanoparticles in the corona of the PMAA-PS-PMAA triblock copolymer, the initial pH value of the aqueous solution was adjusted from 6.8 to 3.2 by adding 0.75mL of 0.1M HCl and from 6.8 to 9.5 by adding 0.75mL of 0.1M NaOH.

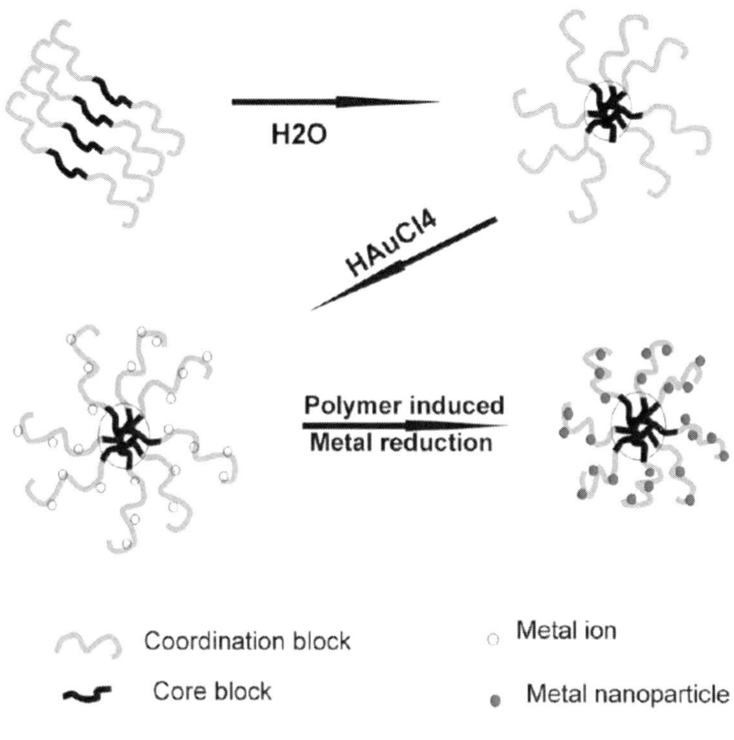

Figure 2. Schematic showing the preparation protocol of block copolymer/ Au metal nanoparticle hybrid aggregates in H_2O.

Characterization Methods

Dynamic light scattering (DLS) was employed in order to investigate the behavior of the copolymer micelles and the hybrid nanosystems. A 0.45μm Millipore filter was used in order to filter all solutions into a dust free cylindrical light scattering cell of 1 cm diameter. All measurements were performed on a ALV/CGS-3 Compact Goniometer System (ALV GmbH Germany) equipped with a JDS Uniphase 22mW He-Ne laser, operating at 632.8 nm, interfaced with a ALV-5000/EPP multi-tau digital corraletor with 288 channels and a LV/LSE-5003 light scattering electronics unit for stepper motor drive and limit switch control. The scattering intensity and correlation functions were measured in the angular range of 45^0 to 135^0. Autocorrelation functions were analyzed by the cumulants method and the CONTIN routine. Apparent hydrodynamic radii were calculated by the Stokes-Einstein equation [23]. All samples were characterized by UV-vis spectroscopy using a Perkin-Elmer UV-vis/NIR Lamda 19 spectrometer. Additional structural information was obtained by transmission electron microscopy (TEM), using a JOEL model JEM100C electron microscope, operated at 80KV accelerating voltage and under bright field conditions. Atomic force microscopy (AFM) images were taken on an Omicron GmbH and SPM Digital Instruments system. Samples for TEM and AFM imaging were prepared by drop casting the hybrid aggregate solutions onto carbon-coated EM copper grids and silicon wafers respectively. Five minutes after the deposition, the excess solution was blotted away with a strip of filter paper, while the rest was allowed to dry at room temperature.

RESULTS AND DISCUSION

As already mentioned water is a selective solvent for the PMAA block, resulting in the formation of triblock micellar aggregates with a PMAA corona and a PS core (where each triblock copolymer chain is assumed to take a U-shape conformation). In order to follow the self-assembling behavior of the micelles and the hybrid aggregates, DLS measurements were performed. In Figure 3, the hydrodynamic radii R_h of two distinct species at 73 and 11 nm are observed after analysis of the intensity correlation function by CONTIN. The one with a hydrodynamic radius of 73 nm should be attributed to triblock copolymer micelles, while the second population observed at around 11 nm is due to the presence of unimers. The existence of equilibrium between micelles and unimers should be attributed to the PS content of the triblock. After the addition of $HAuCl_4$ and the subsequent reduction of Au^{3+} to Au^0, a more compact conformation of the micelles is observed. Once $HAuCl_4$ was added, $AuCl_4^{3-}$ coordinated with the MAA segments at the corona of the micelles and a decrease in the R_h was observed. The reduction of Au^{3+} to Au^0 was presumably facilitated by pseudo macrocyclic domains formed by the PMAA block [24], due to formation of carboxylic acid dimmers, leading to the formation of hybrid micelles incorporating metal nanoparticles. The micelles' R_h decreased to 47nm, while the unimers still existed having the same average size. However their population decreased remarkably, when the gold added molar ratio was 4:1. Probably, the existence of small gold nanoparticles around the triblock copolymer caused an increased incorporation of unimers into the micelles. Nevertheless, the amount of gold added was not enough so as to cause complete micellization of PMAA-PS-PMAA chains and the presence of unimers should not be neglected. When the amount of gold added was increased to MAA:Au 2:1 molar ratio, the micelles showed exactly the same distribution as for the case of MAA:Au 4:1 molar ratio (R_h = 47 nm), however the unimers population decreased futher in size as well as in amount.

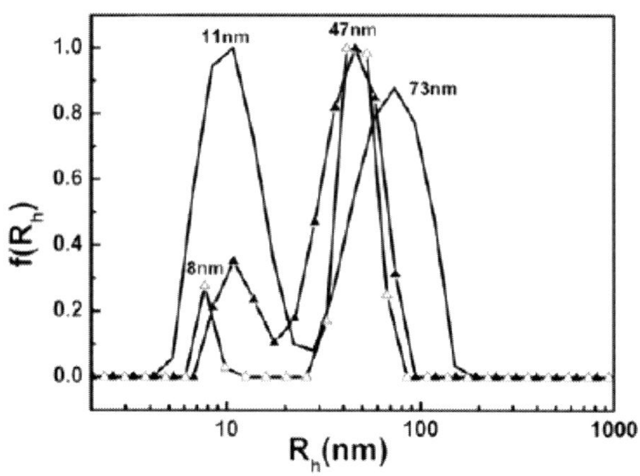

Figure 3. Hydrodynamic radius distribution of PMAA-PS-PMAA in H_2O (solid line), PMAA-PS-PMAA /Au 4:1(closed triangles) and PMAA-PS-PMAA /Au 2:1 (open triangles) complex.

The formation of gold nanoparticles through autoreduction in the presence of aqueous PMAA-PS-PMAA triblock copolymer solutions was followed by absorption spectroscopy. The peak centered at ~546 nm originates from the surface plasmon resonance of gold nanoparticles formed in the triblock copolymer solutions. Figure 4a shows the absorption spectra obtained after mixing the aqueous solution of PMAA-PS-PMAA triblock copolymer with $HAuCl_4$ (MAA:Au 4:1 molar ratio). The spectra were taken every 30min, after gold addition, while heating the system at a moderate temperature (50 ^{o}C), and an increase in the absorption band intensity was observed. This indicated that gold nanoparticles were formed following the reduction of $AuCl_4^-$ which was enhanced due to the increase of the time of thermal treatment. However, the absorption band was remarkably broad implying that either the reduction of Au^{3+} to Au^0 led to the formation of large nanoparticles or that the nanoparticles suffered severe agglomeration due to the thermal treatment, which in turn forced the reduction reaction to proceed faster. The size of the Au NPs formed is estimated to be more than 30 nm [1].

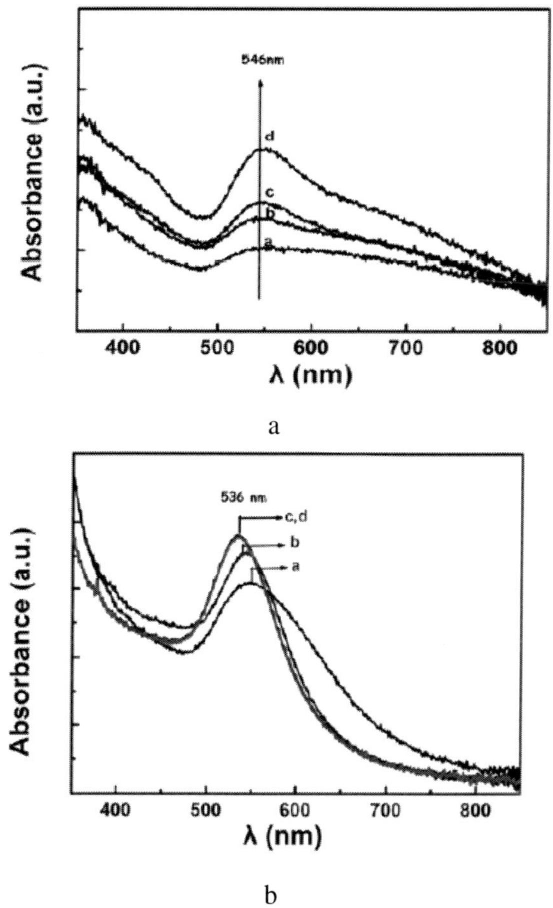

Figure 4. Absorption spectra for PMAA-PS-PMAA /Au ions hybrid complex (4:1 ratio) a) during heating and b) without heating the colloidal system. The arrow indicates the increase in the absorption spectra centered at ~536 nm due to SPR during the gold nanoparticle formation.

In the absence of thermal treatment, only a single well defined and sharp absorption peak was observed indicating that no AuNPs agglomeration occurs, while the size of the Au nanopheres formed should be around 15 nm based on the literature data [1]. However, the reduction reaction is slower, starting at 15hrs after mixing of PMAA-PS-PMAA copolymer and $HAuCl_4$ solutions and taking up to 24hrs until it is completed (as indicated by the SPR band attaining a plateau level). Once the reduction process is completed and the formation of gold nanoparticles is accomplished, the complex PMAA-PS-PMAA/Au nanostructure shows no thermal response and the AuNPs remain stable without forming any aggregates. Increasing the quantity of $HAuCl_4$ added, led to a red shift of the absorption band [13,25]. This SPR shift could be attributed to formation of nanoparticles of greater size, because of their agglomeration, due to excess quantity of gold. Since gold nanoparticles are not trapped inside the micelles, but they exist coordinated to the corona, it is easy to interact with one another and form aggregates.

Further information on the structure of PMAA-PS-PMAA/Au hybrid nanostructure was gathered by transmission electron microscopy (TEM) and atomic force microscopy (AFM). A representative AFM image is shown in Figure 5 for the case of MAA:Au 4:1 ratio. Well defined nanostructures with a nearly spherical shape can be observed on the surface of SiO_2 wafers. The average size of the hybrid PMAA-PS-PMAA/Au nanostructures was ~42 nm as derived by AFM measurements. The average diameter of PMAA-PS-PMAA/Au nanocolloid was found larger by DLS measurements (around ~90 nm). It should be taken into account that DLS measurements are performed in solution of the hybrid PMAA-PS-PMAA/Au nanostructure, thus the solvent plays a key role in the conformation and size of the hybrid nanostructure, where the corona of the micelle is swollen [26], while AFM observations were made in the dry state. Nevertheless, both techniques show the presence of spherical hybrid micelles.

In Figure 6 a TEM image of the PMAA-PS-PMAA/Au micelles is presented, where micelles of average diameter of ~140 nm can be observed. Severe agglomeration of the hybrid micelles is evident probably due to the formation of carboxylic acid dimmers during solvent evaporation on the grid. The contrast in the image should be attributed to the formed gold nanoparticles trapped around the PMAA-PS-PMAA micelles. In this case, the size of the micelles derived by TEM is larger than the one measured by DLS. This disagreement should be attributed to the large aggregation of the primary micelles on the TEM grid, due to the solvent evaporation and the interactions between the micelles and the substrate. Finally, the difference in the sizes measured by AFM and TEM should be attributed to the effective contrast present in the TEM measurements and the different measurement modes of the two techniques. However, the presence of AuNPs is verified by an additional technique, while their size is in the range of 10 nm to 20 nm in accordance to the UV-vis observations[1].

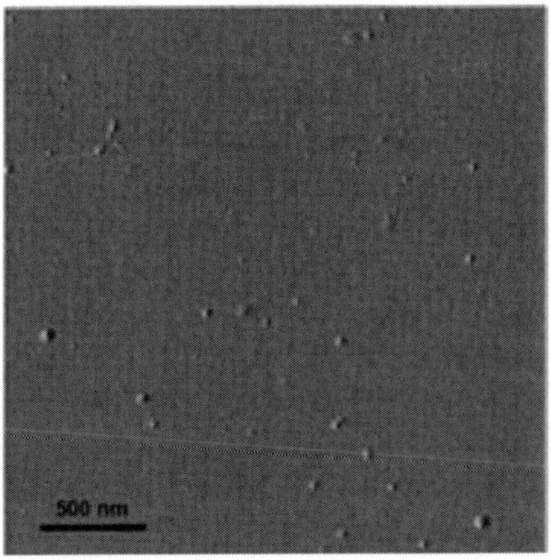

Figure 5. AFM image of PMAA-PS-PMAA/Au nanoparticles hybrid nanostrcuture.

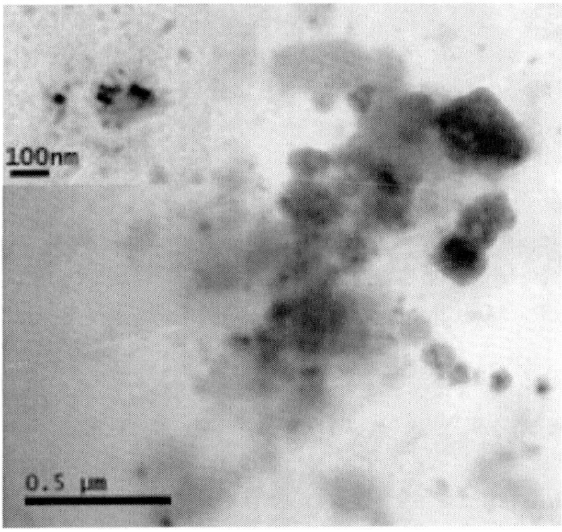

Figure 6. TEM image of PMAA-PS-PMAA/Au nanoparticles hybrid nanostructure.

The dependence of the PMAA-PS-PMAA/Au NPs colloidal nanosystem properties on salt concentration was examined using dynamic light scattering and Uv-vis spectroscopy. These studies provide valuable information for the properties and responsiveness of the nanostructures in conditions met in biological applications (e.g. under physiological conditions where ionic strength is 0.154 M). Various quantities of NaCl were gradually added to the PMAA-PS-PMAA/Au colloidal system (MAA:Au 4:1), obtaining salt concentrations in the range of 0.03M to 0.2M, In order to thoroughly investigate the ionic strength effect on the colloidal nanosystem, three different salt concentrations were studied (0.03M, 0.1M and 0.2M). By DLS measurements, it was observed that the R_h remained almost constant even when 0.2M NaCl were added, although the size distribution was increased (Fig. 7). This could be attributed to the fact that the PMAA-PS-PMAA/Au NPs colloidal material inflates in the presence of NaCl. The carboxylic acid groups of the PMAA block are coordinated with the Na^+ leading to the formation of carboxylic salts. This coordination probably resulted in inflating the colloidal PMAA-PS-PMAA/Au NPs system and at the same time brought a redistribution of the PMAA blocks inside the micellar corona. As observed by UV-vis spectroscopy addition of 0.03M of NaCl led to a prompt increase of the absorption peak and a shift of 6 nm at the absorption maximum (Fig. 8). However, as increasing quantities of NaCl were added, the absorption peak intensity decreased and no further shift of the SPR maximum was observed. The redistribution of the PMAA blocks due to the presence of NaCl, as already mentioned above, presumably resulted in altering the relevant distance between the gold nanoparticles and altered their mutual interactions and immediate environment leading to a red shift of SPR.

The pH effect on the properties of the hybrid PMAA-PS-PMAA/Au NPs nanocolloid was investigated by adjusting the pH of the initial aqueous solutions by the addition of HCl and NaOH respectively, as described in the experimental section. Since all changes regarding pH conditions occur after AuNPs formation, no changes of the size and the shape of the metal nanoparticles are expected. Consequently, any changes observed should be attributed to the influence of the pH to the overall hybrid nanostructure and especially on the PMAA corona chains. In acidic conditions, i.e. at pH 3.2 after addition of 0.1M HCl at the almost neutral (pH 6.8) PMAA-PS-PMAA/Au system, a decrease in the hydrodynamic radius is observed (the R_h decreases to 36 nm from 46 nm initially), without any influence at the size distribution compared to pH 6.8 case. This decrease can be attributed to a more compact conformation that the PMAA chain tends to adopt due to hydrophobic interactions [27], resulting in shrinkage of the hybrid micelle corona, affecting thus the overall size of the nanostructure. A second distribution observed at around ~10 nm could be attributed to the presence of unimers as discussed before. However, the absorption spectra remained unchanged (Figure 10), in terms of peak maximum position.

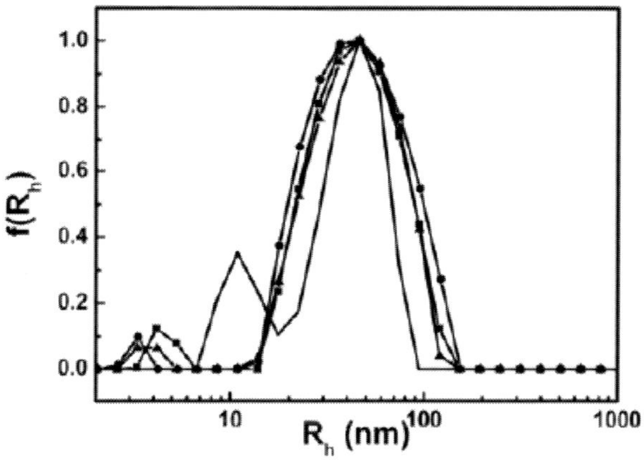

Figure 7. Hydrodynamic radius distribution for PMAA-PS-PMAA/Au NPs 4:1 (solid line), PMAA-PS-PMAA/Au NPs 4:1/NaCl 0.03M (triangles), PMAA-PS-PMAA/Au NPs 4:1/NaCl 0.1 M (squares) and PMAA-PS-PMAA/Au NPs 4:1/NaCl 0.2M (cycles).

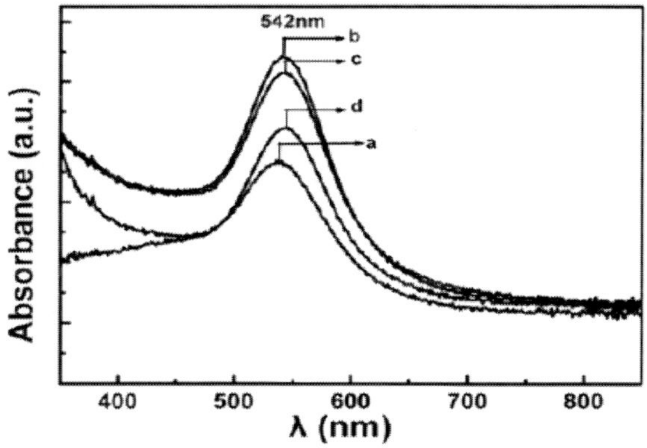

Figure 8. Absorption spectra of a) PMAA-PS-PMAA/Au NPs (4:1) colloid and upon addition of b) 0.03M, c) 0.1M, d) 0.2M NaCl.

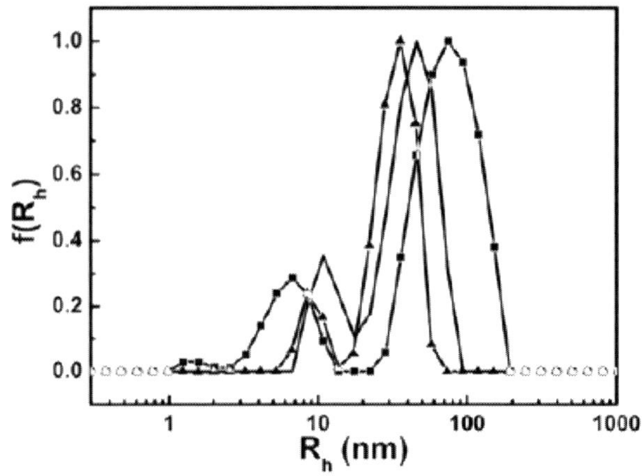

Figure 9. Hydrodynamic radius distribution of PMAA-PS-PMAA/Au NPs (4:1) colloid at pH 6.8 (solid line), pH 3.2 (triangles) and pH 9.5 (squares).

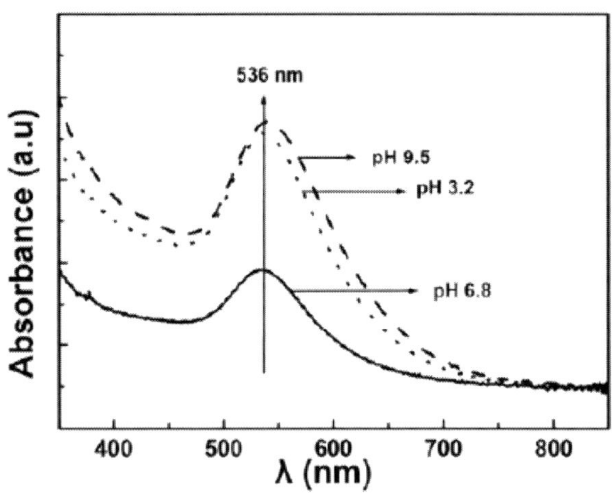

Figure 10. Absorption spectra of PMAA-PS-PMAA/Au NPs (4:1) colloid at pH 6.8 (solid line), pH 3.2 (dotted line) and pH 9.5 (dashed line).

On the other hand, increasing the pH at 9.5 resulted in an increase of the hydrodynamic radius distribution (R_h 75 nm). As the pH increases the carboxylic groups of the PMAA block get more dissociated, resulting in stretching of this block of the copolymer and consequently to an overall increase of the hybrid nanostructure due to the increase of the electrostatic repulsion among the corona chains [28]. A second species with R_h 8 nm was also observed. In these conditions, a blue shift (4 nm) at the position of the UV-vis absorption band was observed (Figure 10), indicating that the relevant distance between neighbouring AuNPs was increased [28], due to the changes imposed within the corona region of the micelles.

CONCLUSION

Au nanoparticles were in situ synthesized through polymer induced reduction within the corona of poly[(methacrylic acid)-b-(styrene)-b-(methacrylic acid)] (PMAA-PS-PMAA) triblock copolymer micelles. The kinetics of this formation was thoroughly studied using dynamic light scattering and UV-vis techniques. It was observed that temperature could accelerate the formation of AuNPs leading however to large agglomerates of metal nanoparticles.

Under alkaline conditions the PMAA-PS-PMAA/Au hybrid system adopts a more loose conformation due to dissociation of the carboxylic groups of the PMAA block, while under acidic conditions a more compact conformation is observed due to hydrophobic interactions. These structural changes are depicted as shifts of the SPR band at the UV-vis spectra of the Au NPs, as well as changes in the SPR intensity.

Finally, regarding the ionic strength effect a more complicated variation of SPR intensity was observed (a sharp initial increase accompanied by shift in the maximum and followed by a decrease with no additional shift) due to a possible redistribution of the PMAA blocks within the corona of the triblock copolymer micelles driven by the formation of carboxylic salts and the screening of electrostatic interactions.

ACKNOWLEDGMENT

The authors would like to thank Dr. D. Achilleos and Prof. M. Vamvakaki for the TEM measurements, and Dr. E. Karoutsos for the AFM measurements.

REFERENCES

[1] Eustis, S; El-Sayed, MA. *Chem. Soc. Rev.,* 2006, 35, 209.

[2] Manea, F; Houillon, FB; Pasquato, L; Scrimin, P. *Angew. Chem. Int. Ed.,* 2004, 43, 6165.

[3] Thomas, KG; Kamat, PV. *Acc. Chem. Res.,* 2003, 36, 888-898.

[4] Filali, M; Meier, MAR; Schuber, US; Gohy, JF. *Langmuir,* 2005, 21, 7995-8000.

[5] Li, J; Shi, L; An, Y; Li, Y; Chen, X; Dong, H. *Polymer,* 2006, 47, 8480-8487.

[6] Alexandridis, P; Tsianou, M. *Europ. Polym. J.*, 2011, 47, 569.

[7] Alexandridis, P; Sakai, T. *Materials Letters*,2006, 60, 1983-1986.

[8] Newman, JDS; Blanchard, GJ. *Langmuir*, 2006, 22, 5882-5887.

[9] Mill, G; Longenbeger, L. *J. Phys. Chem.,* 1995, 99, 475-478.

[10] Sakai, T; Alexandridis, P. *Langmuir,* 2004, *20*, 8426.

[11] Alexandridis, P; Sakai, T. *Chem Mater,* 2006, 18, 2577-1283.

[12] Sabir, TS; Yan, D; Milligan, JR; Aruni, W; Nick, KE; Ramon, RH; Hughes, JA; Chen, Q; Kurti, S; Perry, C.C. *J. Phys. Chem. C,* 2012, 116, 4431.

[13] Alexandridis, P; Sakai, T. *Nanotechnology,* 2005, 16, S344-S353.

[14] Alexandridis, P. *Chem. Eng. Technol.,* 2011, 34, 15.

[15] Khullar, P; Mahal, A; Singh, V; Banipal, TS; Kaur, G; Bakshi, MS. *Langmuir,* 2010, 26, 11363.

[16] Chen, X; Liu, Y; An, Y; Lü, J; Li, J; Xiong, D; Shi, L. *Macromol. Rapid Commun.,* 2007, 28, 1350.

[17] Chen, X; An, Y; Zhao, D; He, Z; Cheng, J; Shi, L. *Langmuir,* 2008, 24, 8198.

[18] Zheng, P; Jiang, X; Zhang, X; Zhang, W; Shi, L. *Langmuir,* 2006, 22, 9393.

[19] Xu, H; Xu, J; Jiang, X; Zhu, Z; Rao, J; Yin, J; Wu, T; Liu, H; Liu, S. *Chem. Mater.,* 2007, 19, 2489.

[20] Zhao, H: Douglas, EP. *Chem. Mater.*, 2002, 14, 418-423.

[21] Hou, G; Zhu, L; Chen, D; Jiang, M. *Macromolecules*, 2007, 40, 2134-2140.

[22] Meristoudi, A; Pispas, S. *Polymer*, 2009, 50, 2743.

[23] Meristoudi, A; Pispas, S; Vainos, N. *J. Polym. Sci. Part B: Polym. Phys.*, 2008, 15, 1515.

[24] Sakai, T; Alexandridis, A. *J. Phys. Chem. B*, 2005, 109, 7766.

[25] Chen, X; An, Y; Zhao, D; He, Z; Cheng, J; Shi, L. *Langmuir*, 2008, 15, 8198-8204.
[26] Pispas; S. J. Polym. Sci. Part A Polym. Chem., 2007, 45, 509-590.
[27] Nuopponen, M; Tenhu, H. *Langmuir*, 2007, 23, 5352.
[28] Li, D; He, Q; Yang, Y; Mo¨hwald, H; Li J. *Macromolecules*, 2008,19, 7254-7256

In: Gold Nanoparticles
Editors: A. Jarnagin and L. Halshauser
ISBN 978-1-62257-927-3

Chapter 5

Review: Application of Gold Nanoparticles for Metabolite Imaging

***Naoko Goto-Inoue*[1,*], *Takahiro Hayasaka*[2] and *Mitsutoshi Setou*[2]**
[1]Department of Health Promotion Sciences, Graduate School of Health Promotion Sciences, Tokyo Metropolitan University, Tokyo, Japan
[2]Department of Cell Biology and Anatomy, Hamamatsu University School of Medicine, Shizuoka, Japan

Abstract

Matrix-assisted laser desorption/ionization (MALDI)-imaging mass spectrometry (IMS) enables visualization of the distribution of metabolites in tissue sections. IMS is a powerful technique for the visualization of endogenous small metabolites, especially lipids, facilitated by the advantages of mass spectrometry-based molecular detection. Despite the promising capability of MALDI-IMS for imaging, there is still some room for improving the technique, especially in spatial resolution and sensitivity.

---

* Corresponding author: Naoko Goto-Inoue; Department of Health Promotion Sciences, Graduate School of Health Promotion Sciences, Tokyo Metropolitan University, 1-1 Minami-Osawa, Hachioji, Tokyo 192-0397, Japan; Tel: (+81) 42 677 2965; E-mail: naoko.goto.inoue@gmail.com.

In this review, we show some applications of gold nanoparticles (AuNPs) for mapping the distribution and localization of small metabolites in tissue sections. We have developed and applied AuNPs as a new matrix to maximize the detection of metabolites in the mouse brain. We believe that our approach using AuNPs in IMS could lead to a new strategy for analyzing basic biological mechanisms and several diseases.

1. Introduction

This review article is devoted to describing applications of gold nanoparticles (AuNPs) in molecular imaging, especially in imaging mass spectrometry (IMS). IMS is a relatively new imaging method based on mass spectrometric analysis onto tissue sections [1]. Several surface analysis methods employ a mass spectrometric approach, such as secondary ion mass spectrometry (SIMS) [2], desorption electrospray ionization mass spectrometry (DESI-MS) [3], or matrix-assisted laser desorption/ionization mass spectrometry (MALDI-MS) [4]. Among the variety of mass spectrometric methods available, MALDI-MS is the most suitable for the global analysis of metabolites because MALDI-MS can detect a wide range of biomolecules ranging in size from small to large [5]. MALDI-IMS is based on two-dimensional MALDI-MS analysis of tissue sections [6]. The versatility of MALDI-IMS has opened a new frontier in several fields, including proteomics [7], lipidomics [8, 9], metabolomics [10], and glycomics [11]. We recently established a method for the visualization of metabolites with AuNPs by MALDI-IMS [12]. In this review, the use of AuNPs for imaging of metabolites will be demonstrated.

2. Chemical Characteristics of Gold Nanoparticles

Nanoparticles have been attracting considerable attention due to their large surface area-to-volume ratio, as well as various distinctive physical and/or chemical properties. For biological applications, nanoparticles are widely used as very small probes that allow researchers to reveal molecular machinery. There are various kinds of nanoparticles with metal cores such as gold [13], silver [14], and titanium dioxide [15]. AuNPs are widely used in immunohistochemistry to identify protein-protein interactions [16].

When gold metal is used as a core on nanoparticles, it exhibits distinctive optical properties due to surface plasmon resonance. Moreover, a recent report demonstrates a key characteristic of AuNPs: that of superior performance in enhancing the sensitivity and/or specificity in measurement sciences. Laser irradiation to AuNPs can induce rapid heating in a small volume of AuNPs. This property can facilitate the efficient desorption/ionization of molecules adsorbed onto AuNPs.

3. Application for Matrix Assisted Laser Desorption/Ionization Mass Spectrometry

MALDI-MS is used widely in various fields because it has the characteristics of analyzing speed, ease of use, high sensitivity, and wide detectable mass range, but suppression effects between analyte molecules and interference from the matrix frequently arise during MALDI analysis. MALDI-MS needs matrices to enhance desorption/ionization of analytes. Small organic compounds are commonly supplied as matrices. Especially, 2,5-dihydroxybenzoic acid (DHB) is used due to its wide range of applicability: small molecular lipids to proteins can be ionized by the use of DHB. It has two disadvantages, however. One is the inadequate homogeneity of the crystal. Crystals of DHB are non-homogeneous, and pointed large crystals are formed on samples. In such conditions, we have to search for a "sweet spot" to get a stable mass spectrum. The other is the ionization suppression of the analyte. DHB itself is easily ionized. Therefore, these DHB-derived ions suppress the ionization of the analytes.

To overcome these problems, the use of an inorganic material, namely AuNPs, as a matrix has been introduced [17, 18]. Mass spectrometry using NPs differs from spectrometry using organic matrices in several ways. First, the particle size of AuNPs is easily controlled, and homogeneous crystals are easily prepared on samples. Second, the mass spectra are simplified owing to a lack of chemical noise found in organic matrices [17]. Therefore, compared with organic chemical matrices, AuNPs offer the advantages of eliminating matrix ion interference and improving sample homogeneity.

Recently, size-selected (2-10 nm) AuNPs have been used as a matrix in laser desorption/ionization mass spectrometry, enabling the selective and sensitive detection of phosphopeptides [13]. Moreover, their application in MALDI has also been demonstrated [19]. Wu et al. demonstrated that AuNP inorganic matrices possess high salt tolerance for the analysis of biomolecules, and they succeeded in using it to ionize small metabolites such as progesterone [20]. AuNPs facilitate the property of efficient desorption/ionization of molecules, and this tendency shows a specificity to carbohydrates which is hardly detected with organic matrices without derivatization. According to these characteristics, the determination of neutral carbohydrates has been accomplished by AuNP-assisted LDI [21]. With this approach, the ionization efficiency of neutral carbohydrates can be greatly amplified without derivatization steps.

4. Imaging Mass Spectrometry with Gold Nanoparticles

The advantages of IMS are that it can visualize the distribution of biomolecules without derivatization and labeling and can identify the detailed structures of biomolecules on tissue sections. IMS analyses using the conventional matrix, DHB, have been developed to reveal the distribution of biomolecules such as phospholipids [8], proteins [22], and glycosphingolipids (GSLs) [11, 23]. In IMS analyses, a homogeneous matrix coating is necessary to get appropriate molecular distribution on tissue sections. There are 3 commonly used matrix application methods: spraying, depositing, and sublimation. They all require some specific skills or machines, and the crystal size sometimes exceeds the size of one cell. Therefore, the advantage of AuNPs, small crystals and homogeneous coating of the samples is very useful in IMS analyses.

It is demonstrated that the mass spectrum acquired with AuNPs was quite different from that with organic matrices since AuNPs removed significant interference from ionized gold clusters and their attachments and allowed spot profile analysis of weak signals from peptides and minority lipid species in the presence of the dominant majority lipid signals [24]. Motivated by these prior results, Jackson et al. have used size-controlled AuNPs to measure the spatial distribution of cerebroside species in rat brain sections. Major mass peaks detected in this spectrum correspond to sodiated and potassiated cerebrosides, which are hard to detect with organic matrices. Tang et al. demonstrated latent fingerprint (LFP) detection for analyzing endogenous and exogenous compounds embedded in LFPs and imaging their distribution without disturbing the finger print patterns [25].

Here we show our application of AuNPs for imaging of sulfatides on mouse brain tissue. We applied AuNPs by spraying onto tissue sections. Our AuNPs, with a mean diameter of 4.3 ± 0.7 nm, exhibited metal content as high as 97%, and could be prepared from hexane solution without additives. They were easy to spray and formed homogeneous crystals.

Sulfatide is an abundant glycosphingolipid component of both the central and the peripheral nervous system and is an essential lipid component of myelin membranes. Sulfatide is the collective name for glycosphingolipids with a polar head consisting of galactosyl-3-*O*-sulfate but with variation in the fatty acids in the ceramide portion.

In a previous study 10 kinds of sulfatides in the mouse brain were detected with a DHB matrix [26], though it was impossible to visualize all sulfatide species by IMS analyses because of the low signals. Figure 1 presents the mass spectra directly acquired from brain sections with DHB (Figure 1a) and AuNPs (Figure 1b).

We could detect *m/z* 888.5 as the major peak in both; however, the signal intensity at *m/z* 888.5 with DHB was 206 while that with AuNPs was 4000. The increased intensity of the sulfatide peak suggests that AuNPs constitute a suitable matrix for the visualization of sulfatides. Each intensity with AuNPs was approximately 20 times higher than that with DHB. We selected 10 sulfatides and compared their ion images by DHB and the AuNPs matrices (Figure 1c and d).

We set the maximum intensity at 1000 or 200 to construct all ion images. With the DHB matrix, we could hardly reveal the distributions for ion images at *m/z* 822.5, 834.5, 850.5, 862.5, 914.5, 916.5, 932.5, and 934.5. On the contrary, we could reliably obtain ion images of all of these ions with AuNPs.

Next, we attempted to show the application of AuNPs for visualization of gangliosides. Gangliosides consist of an oligosaccharide head group containing at least one sialic acid residue. This class of compounds presents a challenge for MS analysis because of the ready elimination of sialic acid [27]. As we conducted MS analysis of gangliosides using AuNPs, we could observe minor components of gangliosides (GM3, GM2, and GD3) as well as the major components (GM1 and GD1) (Figure 2).

All of these detected gangliosides have two kinds of molecular species owing to the difference of the sphingosine (d18:1 or d20:1).

The GM1 and GD1 (d18:1/C18:0) appeared clearly in the cortex and hippocampus. On the other hand, the GM1 and GD1 (d20:1/C18:0) mappings were slightly different from results obtained for d18:1/C18:0. These results confirmed recent studies, which showed that GM1 and GD1 with d20:1 are mainly located in the molecular layer of the dentate gyrus in the mouse brain hippocampus [23]. We could realize a tendency for different distributions depending on the sphingoid base (d18:1 or d20:1) with GM1, GD1, and GD3. Therefore our results demonstrated that the use of AuNPs as a matrix improved the detection sensitivity for glycosphingolipids.

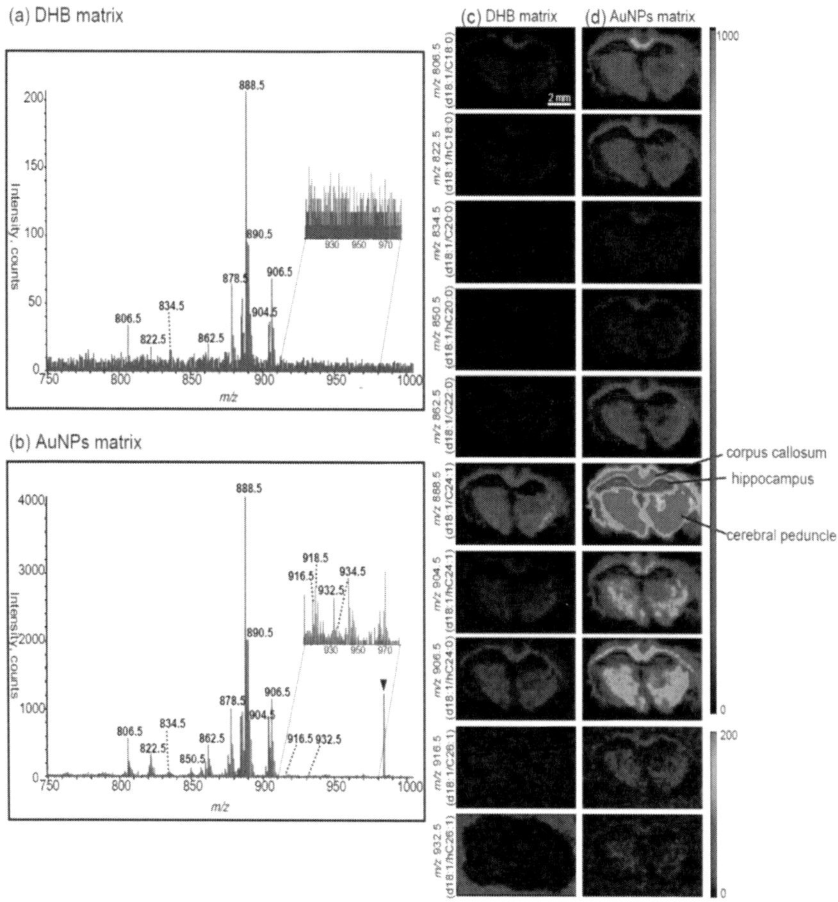

Figure 1. Negative-ion-mode MALDI imaging mass spectrum acquired from mouse brain tissue with DHB (a) and with AuNPs (b). The black triangle shows the peak derived from AuNPs. Ion images of 10 kinds of sulfatide molecular species with a DHB matrix (**c**) and AuNP matrix (d). (Permission from [12]).

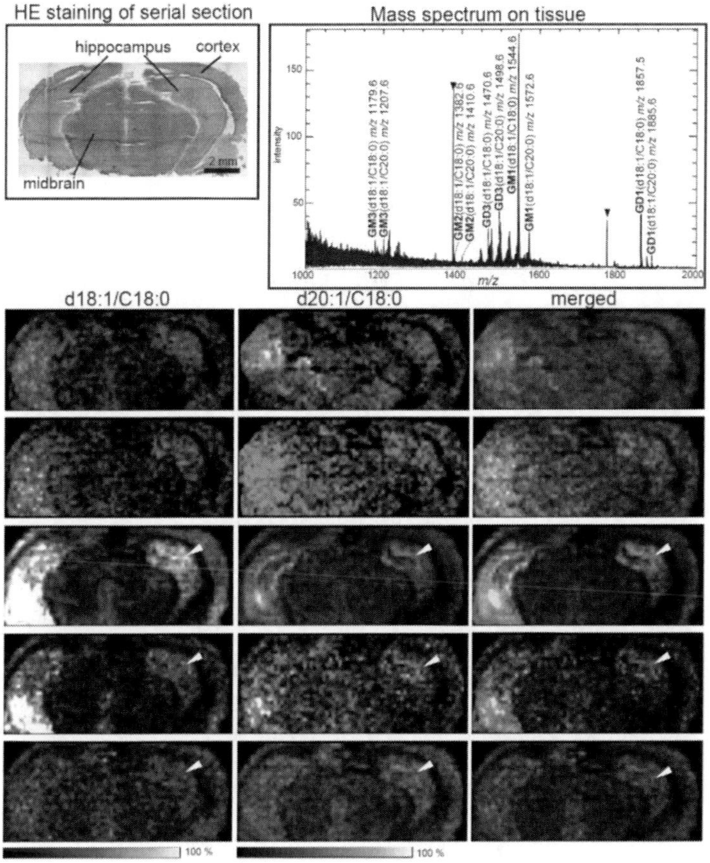

Figure 2. IMS analyses of gangliosides using AuNPs in negative-ion mode. The brain anatomy described in the HE staining result. Mass spectrum on tissue and different ion images for GM3, GM2, GM1, GD1, and GD3, depending on the sphingosine (d18:1 or d20:1), are shown. The black triangle on the mass spectrum shows the peak derived from the AuNPs. White arrows show the localization of the molecular layer of the dentate gyrus in the mouse brain hippocampus. (Permission from [12]).

CONCLUSION

MALDI-IMS analysis has the potential to contribute greatly to molecular imaging studies. AuNPs which utilize higher sensitivity and spatial resolution are among the best materials for MALDI-IMS. Modifications of the features of AuNPs such as side-chains or particle size are expected to increase both the range of detectable molecules and the fields in which AuNPs can be used. In brief, the use of AuNPs is expected to expand the applicability of IMS in the near future.

References

[1] Chughtai K, Heeren R M: Mass spectrometric imaging for biomedical tissue analysis. *Chem Rev*, 2010, 110(5): 3237-77.

[2] Benninghoven A, Sichtermann W K: Detection, identification and structural investigation of biologically important compounds by secondary ion mass spectrometry. *Analytical Chemistry*, 1978, 50(8): 1180-4.

[3] Takats Z, Wiseman J M, Gologan B, Cooks R G: Mass spectrometry sampling under ambient conditions with desorption electrospray ionization. *Science*, 2004, 306(5695): 471-3.

[4] Karas M, Hillenkamp F: Laser desorption ionization of proteins with molecular masses exceeding 10,000 daltons. *Analytical Chemistry*, 1988, 60(20): 2299-301.

[5] Yates J R, 3rd: Mass spectrometry and the age of the proteome. *Journal of Mass Spectrometry*, 1998, 33(1): 1-19.

[6] Stoeckli M, Chaurand P, Hallahan D E, Caprioli R M: Imaging mass spectrometry: a new technology for the analysis of protein expression in mammalian tissues. *Nat Med*, 2001, 7(4): 493-6.

[7] Morita Y, Ikegami K, Goto-Inoue N, Hayasaka T, Zaima N, Tanaka H, Uehara T, Setoguchi T, Sakaguchi T, Igarashi H, Sugimura H, Setou M, Konno H: Imaging mass spectrometry of gastric carcinoma in formalin-fixed paraffin-embedded tissue microarray. *Cancer Science*, 2010, 101(1): 267-73.

[8] Hayasaka T, Goto-Inoue N, Zaima N, Kimura Y, Setou M: Organ-specific distributions of lysophosphatidylcholine and triacylglycerol in mouse embryo. *Lipids*, 2009, 44(9): 837-48.

[9] Zaima N, Hayasaka T, Goto-Inoue N, Setou M: Matrix-assisted laser desorption/ionization imaging mass spectrometry. *Int J Mol Sci*, 2010, 11(12): 5040-55.

[10] Zaima N, Hayasaka T, Goto-Inoue N, Setou M: Imaging of metabolites by MALDI mass spectrometry. *Journal of Oleo Science*, 2009, 58(8): 415-9.

[11] Goto-Inoue N, Hayasaka T, Zaima N, Setou M: The specific localization of seminolipid molecular species on mouse testis during testicular maturation revealed by imaging mass spectrometry. *Glycobiology*, 2009, 19(9): 950-7.

[12] Goto-Inoue N, Hayasaka T, Zaima N, Kashiwagi Y, Yamamoto M, Nakamoto M, Setou M: The detection of glycosphingolipids in brain tissue sections by imaging mass spectrometry using gold nanoparticles. *J Am Soc Mass Spectrom*, 2010, 21(11): 1940-3.

[13] McLean J A, Stumpo K A, Russell D H: Size-selected (2-10 nm) gold nanoparticles for matrix assisted laser desorption ionization of peptides. *J Am Chem Soc*, 2005, 127(15): 5304-5.

[14] Hayasaka T, Goto-Inoue N, Zaima N, Shrivas K, Kashiwagi Y, Yamamoto M, Nakamoto M, Setou M: Imaging mass spectrometry with silver nanoparticles reveals the distribution of fatty acids in mouse retinal sections. *J Am Soc Mass Spectrom*, 2010, 21(8): 1446-54.

[15] Shrivas K, Hayasaka T, Sugiura Y, Setou M: Method for simultaneous imaging of endogenous low molecular weight metabolites in mouse brain using TiO2 nanoparticles in nanoparticle-assisted laser desorption/ionization-imaging mass spectrometry. *Analytical Chemistry*, 2011, 83(19): 7283-9.

[16] Sanvicens N, Marco M P: Multifunctional nanoparticles--properties and prospects for their use in human medicine. *Trends Biotechnol*, 2008, 26(8): 425-33.

[17] Chen L C, Yonehama J, Ueda T, Hori H, Hiraoka K: Visible-laser desorption/ionization on gold nanostructures. *Journal of Mass Spectrometry*, 2007, 42(3): 346-53.

[18] Peterson D S: Matrix-free methods for laser desorption/ionization mass spectrometry. *Mass Spectrom Rev*, 2007, 26(1): 19-34.

[19] Kim Y P, Oh E, Hong M Y, Lee D, Han M K, Shon H K, Moon D W, Kim H S, Lee T G: Gold nanoparticle-enhanced secondary ion mass spectrometry imaging of peptides on self-assembled monolayers. *Analytical Chemistry*, 2006, 78(6): 1913-20.

[20] Wu H P, Yu C J, Lin C Y, Lin Y H, Tseng W L: Gold nanoparticles as assisted matrices for the detection of biomolecules in a high-salt solution through laser desorption/ionization mass spectrometry. *J Am Soc Mass Spectrom*, 2009, 20(5): 875-82.

[21] Jackson S N, Ugarov M, Egan T, Post J D, Langlais D, Albert Schultz J, Woods A S: MALDI-ion mobility-TOFMS imaging of lipids in rat brain tissue. *Journal of Mass Spectrometry*, 2007, 42(8): 1093-8.

[22] Yao I, Sugiura Y, Matsumoto M, Setou M: In situ proteomics with imaging mass spectrometry and principal component analysis in the Scrapper-knockout mouse brain. *Proteomics*, 2008, 8(18): 3692-701.

[23] Sugiura Y, Shimma S, Konishi Y, Yamada M K, Setou M: Imaging mass spectrometry technology and application on ganglioside study; visualization of age-dependent accumulation of C20-ganglioside molecular species in the mouse hippocampus. *Plos One*, 2008, 3(9): e3232.

[24] Novikov A, Caroff M, Della-Negra S, Lebeyec Y, Pautrat M, Schultz J A, Tempez A, Wang H Y, Jackson S N, Woods A S: Matrix-implanted laser desorption/ionization mass spectrometry. *Analytical Chemistry*, 2004, 76(24): 7288-93.

[25] Tang H W, Lu W, Che C M, Ng K M: Gold nanoparticles and imaging mass spectrometry: double imaging of latent fingerprints. *Analytical Chemistry*, 2010, 82(5): 1589-93.

[26] Groseclose M R, Andersson M, Hardesty W M, Caprioli R M: Identification of proteins directly from tissue: in situ tryptic digestions coupled with imaging mass spectrometry. *Journal of Mass Spectrometry*, 2007, 42(2): 254-62.

[27] Harvey D J: Analysis of carbohydrates and glycoconjugates by matrix-assisted laser desorption/ionization mass spectrometry: An update covering the period 1999-2000. *Mass Spectrom Rev*, 2006, 25(4): 595-662.

In: Gold Nanoparticles ISBN 978-1-62257-927-3
Editors: A. Jarnagin and L. Halshauser

Chapter 6

INTRACELLULAR DELIVERY OF GOLD NANOPARTICLES: APPLICATIONS IN NANOMEDICINE

Deendayal Mandal *

Department of Biomedical and Pharmaceutical Sciences,
University of Rhode Island, Kingston, RI, US

ABSTRACT

Nanotechnology has enormous potential in biomedical research including disease diagnosis and therapy. Inorganic nanoparticles have been adopted for biomedical research due to their unique optical and physical properties. Compared to conventional materials, inorganic nanomaterials have several advantages such as simple preparative process and precise control over their shape, composition and size. A major aim of medicine is the early and precise diagnosis of clinical conditions, providing an efficient treatment without secondary effects. Nanoparticle-based technologies have demonstrated especially high potential for medical purposes, ranging from diagnosing diseases to providing novel therapies. Among the metal nanoparticles, gold nanoparticles have proven to be powerful tools in various nanomedicinal applications because of their photo-optical properties and biocompatibility. However, the existing

* Corresponding author: Department of Biomedical and Pharmaceutical Sciences, 41 Lower College Road, University of Rhode Island, Kingston, RI 02881 (USA), Email: ddmandal@yahoo.com; Ph: +1 401 874 4221;, Fax: +1 401 874 5787.

nanoparticle-based technologies must overcome several challenges, including selective nanoparticle delivery, potential toxicity, imaging of nanoparticles and therapeutic efficacy. This chapter describes a brief overview of the recent research on delivery of gold nanoparticles inside the cells and their potential applications in nanomedicine.

INTRODUCTION

Targeted entry into cells is an important area of research in drug delivery and therapy [1-4]. Site-specific delivery of drugs and therapeutics is essential to reduce drug toxicity and increase therapeutic effects. To achieve efficient and specific cellular drug delivery, one strategy is to modify or conjugate drug with a ligand that can be efficiently taken up by target cells via receptor-mediated endocytosis [5].

The application of nanoparticles to nanomedicine research has tremendous potential in biomedical research, disease diagnosis and therapy. Nanoparticles has drawn attraction because of their interesting surface properties. Most of the applications are based on the optical properties of nanoparticles such as surface Plasmon resonance, surface enhanced Raman scattering and strong photoluminescence. Among the metal nanoparticles gold nanoparticles have been used with a long history as a contrast agent for scanning electron and light microscopy in histopathology study [6]. Some complete review articles have recently been published on the biological applications of metal nanoparticles. [7, 8]. Cellular internalization of nanoparticles that have optical properties is very important in the cellular imaging and targeted therapy. However, intracellular delivery of metal nanoparticles is a challenging task as the cell membrane is a formidable barrier to foreign materials [9-11]. The surface modification of nanoparticles is important to overcome this limitation. Surface modifications of metal nanoparticles with biomolecules (peptides, aptamers, antibodies, and other molecules) has facilitated specific interactions of nanoparticles with target cells or organelles [12-16].

Colloidal gold nanoparticles are one of the attractive platforms for cancer diagnosis and therapy. Gold nanoparticles are attractive because it has been approved and used for treatment of human disease (e.g. rheumatoid arthritis) [17]. Gold nanoparticles are also attractive because they are relatively easy to synthesize. For efficient delivery of AuNPs into a living system it needs to overcome natural biological barriers such as the cell membrane and the reticuloendothelial system (RES). Gold makes an excellent intracellular targeting vector for three reasons. First, gold can be synthesized routinely in sizes varying continuously from 0.8 to 200 nm with <10% size dispersity. Second, gold can be modified with a large collection of small molecules, peptides, proteins, DNA, and polymers. Moreover, all of these functional elements can be combined on a single particle simultaneously, often via simple one-pot procedures. Finally, gold particles have enormous visible light extinction characteristics.

SYNTHESIS OF AUNPS

Several methods have been reported in the literature for the synthesis of AuNPs. The most common method is by chemical reduction of gold salts such as hydrogen tetrachlorate ($HAuCl_4$) using citrate [18] or borohydride [19, 20] as the reducing agent. To achieve better monodispersity Brown and Natan [21] have reported the synthesis of monodisperse AuNPs with diameters between 30 and 100 nm using a seeding approach. The method is based on the use of the surface of the AuNPs as a catalyst for the reduction of Au^{3+} by hydroxylamine. Later this method have been used to control the shape and size of the nanoparticles [22].

There are also several physical methods reported for the synthesis of AuNPs such as microwave irradiation [23-25], sonochemical method [26-29], UV irradiation [30-32], laser ablation [33, 34], thermolysis [35, 36], radicals induced method [37-39], and photochemical method [40].

Biological systems, masters of ambient condition chemistry, synthesize inorganic materials that are hierarchically organized from the nano- to the macroscale. Recent studies on the use of microorganisms in the synthesis of nanoparticles are a relatively new and exciting area of research with considerable potential for development [41]. Intracellular and extracellular gold nanoparticles can be synthesized using fungi [42, 43].

PROTEIN MEDIATED DELIVERY

Several groups have shown that large AuNPs are quickly opsonized by blood and eliminated by the reticulo endothelial system (RES) in mammalian cells [44-47]. Several research groups have grafted different biological macromolecules such as proteins, peptides or small molecules onto AuNPs to attempt cellular selectivity, internalization and localization within heterogeneous population of cancer cells in solid tumors. Proteins such as tumor necrosis factor –α (TNFα) and anti-EGFR antibodies have been successfully grafted onto AuNPs surface and utilized in conjugation with hyperthermia to selectively kill cancer cells. Tumor necrosis factor –α (TNFα) is a potent cytokine with anticancer efficacy. Additionally, TNFα is overexpressed in solid tumors. TNFα grafted onto AuNPs surface has reduced systemic toxicity compared to the native unconjugated TNFα. More importantly, the TNFα-AuNPs conjugate are able to accumulate preferentially in the tumor vasculature. Recently, one chemotherapy agent, Aurimune has been developed by CytImmune (Rockville, MD, USA) based on this technology. Aurimune or CYT-6091 is a multivalent drug that is assembled on 26-nm particles of colloidal gold and designed to deliver recombinant human TNF within solid tumors. The drug is manufactured by covalently linking molecules of TNF and thiol-derivatized polyethylene glycol (PEG-THIOL or PT) onto the surface of the colloidal gold nanoparticles [48]. PEG is used to avoid RES. They have shown that the drug is internalized into MC-38 tumor cells through TNFα-mediated RME. In combination with hyperthermia this multivalent drug is able to kill tumor cells via vascular damage. The survival of SCK mammary carcinoma cells assessed using *in vivo/ in vitro* cell survival assay following treatment with PT-Au- TNFα and hyperthermia alone or combined. The combination treatment of SCK tumors *in vivo* reduced the *in vivo/in vitro* tumor cell survival to 0.05% immediately following heating at 42.5 C and to 0.005% at 18 hrs after heating. [49]. Most probably, this combined effect cause shutdown of tumor microvasculature and microvasculature and thus depriving the tumor microenvironment of nutrients and oxygen [50, 51]. Apparently, the microvascular shutdown makes the tumor tissue more susceptible to thermal injury, and large numbers of tumor cells were cut off from the circulation causing massive cell death and tumor regression.

Another receptor that can be targeted in therapy is epithermal growth factor receptor (EGFR) which is overexpressed in several cancers [52-57]. Monoclonal antibodies such as anti-EGFR antibody have been investigated as a possible anticancer therapy for lung cancer. AuNPs have been successfully conjugated with anti-EGFR antibody and employed as a cancer diagnostic tool [58] and for phtotothermal therapy [59, 60] in an oral cancer model. The color scattering property of the AuNPs has been used in designing anti-EGFR-AuNPs conjugates for diagnosis of HSC and HOC oral cancer. These conjugates bind to cancerous and noncancerous cells in a different manner. In case of HOC and HSC oral cancer cells where EGFR is overexpressed, these conjugates bind in a homogenous manner, whereas HaCaT, a noncancerous cell line in which EGFR expression is depressed, only showed a random AuNPs conjugate binding. Such binding preferences and unique light scattering property of thsese conjugates provide a useful diagnostic tool to distinguish between noncancerous and cancerous cells. This concept has been successfully employed in imaging EGFR-enriched Tu696 human head-and-neck carcinoma cells *in vivo and in vitro* using an ScFr version of the anti-EGFR antibody [61].

Transferrin (TF) is also a suitable ligand for conjugating drugs, since it can be specifically recognized and taken up by TF receptors actively expressed on the surface of various tumors cells [62]. AuNPs-TF conjugation has been made to study the cellular uptake [63]. AuNPs are covalently coupled to TF molecules by mercaptoacetic acid. When the AuNPs-TF conjugates are incubated with human nasopharyngeal carcinoma cells (NPC, SUNE1), nanoparticles internalization into cells is visualized by a multimode Atomic Force Microscopy (AFM). The nanoparticle internalization into NPC cells is validated by using confocal laser scanning microscope (Figure 1). It has been demonstrated that endocytosis of TF-conjugated nanoparticles taking place during the process of internalization. Li et al have recently extended this idea in cancer cell imaging and therapy [64]. Gold nanoparticles are conjugated with transferrin molecules (Tf) for targeting, imaging and therapy of breast cancer cells (Hs578T). They have shown that, the transferrin-transferrin receptor-mediated cellular uptake of gold nanoparticles is six times of that in the absence of this interaction. To demonstrate the efficiency of the conjugates in selectively targeting cancer cells, they have also investigated normal cells. The cellular uptake of the conjugates by the normal cells (3T3) is only one fourth of that by cancerous cells. It is observed that dark field light scattering from cancer cells was enhanced when the cells are incubated with PEG-Tf-AuNPs and very weak scattering without nanoparticles. Photothermal therapy is carried out for cancer cells. It is shown that the cells without AuNPs could not be killed by laser even at a very high power density of 114 W/cm^2 due to the weak light absortion of the cells. However, when the cells are incubated with PEG-AuNPs and PEG-Tf-AuNPs, significant damage to the cells is observed at 7 W/cm^2 in presence of later conjugates. This could be due to the presence of more gold nanoparticles inside cells as a result of specific transferring receptor-mediated cellular uptake, as suggested by the authors. The result demonstrates that this conjugate can be potentially used for cancer diagnostics and therapy.

Peptide Mediated Delivery

Peptides are also good vehicles for AuNP delivery. Most peptides used for AuNPs delivery target the cell nucleus. The nucleus is an important target for photothermal therapy because it contains the cellular genetic machinery. These nuclear membrane-penetrating peptides facilitate the entry of AuNPs into the nuclei by nuclear localization through interaction with the nuclear pore complex [65]. Most of the nuclear membrane-penetrating peptides are derived from virus sources, such as Simian virus [65-67], adenovirus [65], HIV1 [68]. Some RGD peptides have been used to deliver AuNPs [69, 70]. Targeted nuclear delivery is a challenging task, however, as any cell-specific nuclear probe must satisfy the following requirements: it must (i) be small enough to enter cells and cross the nuclear membrane (<100 nm for uptake by receptor-mediated endocytosis (RME) and <30 nm for import through nuclear pores), (ii) penetrate cellular membranes or bind to cell-specific plasma membrane receptors, (iii) bypass or escape endosomal/lysosomal pathways, (iv) penetrate nuclear membranes or access importins to pass through the nuclear pore complex, and (v) have low toxicity.

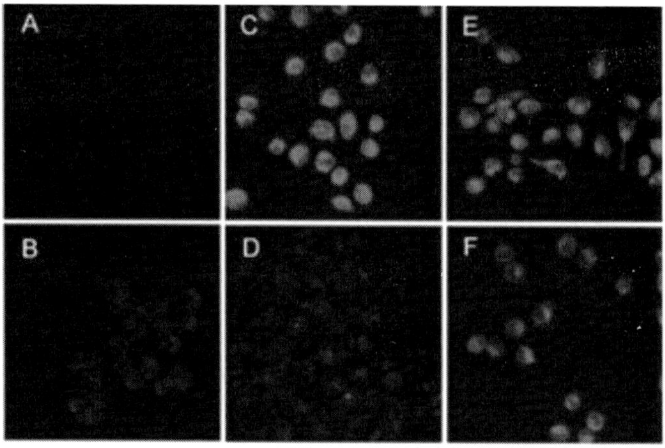

Figure 1. Confocal cell images show that Au-TF nanoparticles were internalized into the cellular plasma of NPC cells via Tfreceptor interaction pathway (5-h treatment). (A) Control cells without Au-TF nanoparticles; (B) cells treated with Au nanoparticles alone; (C) cells incubating with Au-TF nanoparticles; (D) cells treated with Au-albumin nanoparticles; (E and F) cells cotreated with different proportions of Au-TF verses holo-TF (1:2 and 1:5, respectively).

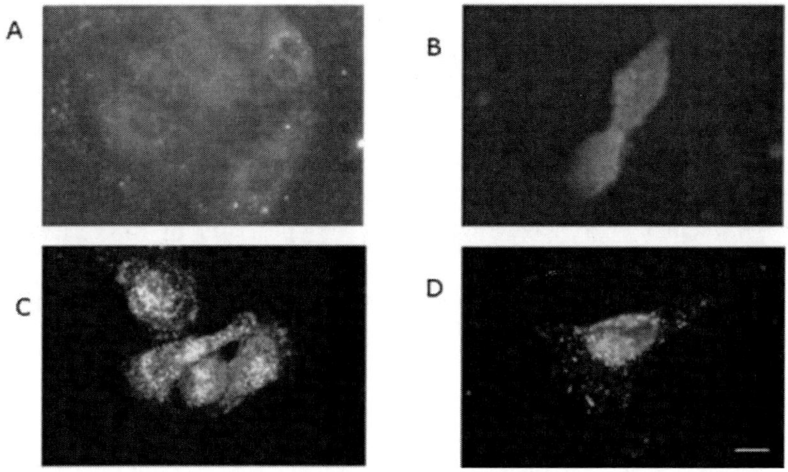

Figure 2. Dark field light scattering images of cells after incubation with gold nanorods and peptide-conjugated gold nanorods for 2 h. A. HaCat normal cells incubated with gold nanorods. B. HSC cancer cells incubated with gold nanorods. C. HaCat normal cells incubated with peptide-conjugated gold nanorods. D. HSC cancer cells incubated with peptide-conjugated gold nanorods. Peptide conjugation promotes the cellular uptake of gold nanorods. Scale bar: 10 μm.

Among the nuclear penetrating peptides identified so far, simian virus nuclear localization peptides (NLS) is very well-known to taget nucleus. Several goups have used NLS peptides to deliver AuNPs inside cells [65, 69, 71]. NLS peptides successfully facilitates the entrance of the conjugate into the nucleus of HepG2 when directly injected inside the cytoplasm [65, 72]. However, the conjugates are trapped inside the cytoplasm when included in the cell growth media. Feldherr et al have extended their study by using different size of gold nanoparticles in simian virus 40 (SV 40) transformed cells [73]. Transformation is dependent on the expression of the early region of the SV40 genome that encodes the large (T) and small (t) tumor antigen [74, 75] . T antigen is a multifunctional protein that conatins a NLS . The experimental cells are microinjected with nucleoplasmin-coated gold particles that ranged in diameter from 110 to 320 A^0, including the 30 A^0 protein coat. Nucleoplasmin is a karyophilic *Xenopus oocyte* protein that conatins a well characterized NLS [76] and can effectively target gold particles to the nucleus [77]. It is observed that nuclear uptake was significantly higher in the transformed cells. In addition, measurements of particle size reveals a significantly increased nuclear accumulation of gold particles in the 200 to 240

A size class. Particles larger than 280 A (or 310 A including protein coat) are unable to enter the nuclei. These results demonstrate that the exclusion limit for particles entering the nucleus was 40 A greater in transformed cells compared with nontransformed cells. In these studies the nanoparticle-peptide complex was injected directly into the cytoplasm near the nucleus, thus bypassing cellular membrane entry and and endosomal /lysosomal pathways. Later Tkachenko et al have investigated on the basis of this established procedure to examine the ability of a number of viral targeting peptides to target the nucleus from outside intact cells [69]. Gold-peptides conjugates are evaluated for their subcellular distribution in HeLa human cervical epithelium cells, 3T3/NIH murine fibroblastoma cells, and HepG2 human hepatocarcinoma cells. Video-enhanced color differential interference contrast microscopy and transmission electron microscopy indicates that transport of nanoparticles into the cytoplasm and nucleus depends on peptide sequence and cell line. In case of HeLa cells, peptide-nanoparticle conjugates are found clustered together in cytoplasm and also few located inside the nucleus. None of the conjugates are able to translocate the nuclear membrane in 3T3/NIH cells. Entry into the cytoplasm of HepG2 cells is strongly dependent on the peptide sequence attached to nanoparticles. The conjugates are able to enter the cytoplasm but were not found near the nuclear membrane or inside the nucleus, which suggests that these conjugates are unable to escape from endosomes. The SV40 NLS is capable of entering HeLa cells via receptor-mediated endocytosis. It is also able to cross the nuclear membrane if it is directly injected into the cytoplasm [73]. Regardless of this, the SV40 NLS is not translocated from outside an intact HeLa cell to the nucleus because it cannot break free of endosomal compartments or is somehow altered as it travels through the endosomal pathway. Gold nanosperes used in these conjugates only absorb in the visible region, thus limiting their use in *in vitro* applications such as in photothermal therapy, which requires near-infrared region. To address this issue, Oyelere et al have used gold nanorods and made AuNPs-NLS peptide conjugates through a thioalkyl-triazole linker [67]. Cellular uptake of CTAB-capped gold nanorods and thioalkyl-triazole-peptide conjugated gold nanorods has been studied. Peptide conjugated gold nanorods have higher efficiency compared with CTAB-capped gold nanorods (Figure 2). This enhanced gold nanorod uptake may be due to the receptor-mediated endocytosis. Moreover, it is found that the thioalkyl-triazole-peptide conjugated nanorods are distributed into both nucleus and cytoplasm in either normal or cancer cells. However, the nanorods are more concentrated in the nucleus of the cancer cells. This may be due to the increased specific and active uptake of the NLS peptide or decreased regulation of passive uptake of cytoplasmic components due to the disruption of the normal cellular process. This group also discovers that Raman signal distinguishes the malignant from nonmalignant cells. This work demonstrates that gold nanorods can be used for nuclear targeting as a new type of imaging-based contrast agents.

Peptides derived from adenovirus have also been studied to promote nuclear penetration [65, 69]. The protein sequence from adenovirus contains both RME and NLS domains [65]. The AuNPs conjugates of this peptide has been shown to successfully avoid the endosomes and penetrates the nucleus of the HepG2 cells [65, 69]. The ability of individual RME and NLS domains derived from the adenovirus protein has been investigated separately for nuclear penetration. Adenoviral RME only able to deliver AuNPs into cytoplasm. NLS sequence of adenovirus protein is unable to enter into HepG2 cells [65], however it delivers AuNPs into the cytoplasm of 3T3/NIH cells and some nuclear translocation has been observed in HeLa cells [69]. Nonetheless, combination of adenoviral RME and NLS conjugated to AuNPs is able to penetrate the nucleus of HepG2 cells. These conjugates show preferential nuclear entry in comparision to the adenovirus protein that contains both RME and NLS sequence [65, 78].

Another nuclear penetrating peptide is derived from HIV1 Tat-protein . This peptide sequence has been used as an efficient way of internalizing a number of marker proteins in cells [79, 80]. Gold nanoparticles are conjugated with Tat peptide via a non-amino acid linker, tiopronin [68]. These nanoparticles are biocompatible. Both Au-tiopronin and Au-Tat conjugates have been studied for cellular uptake. These conjugates are taken up by the primary human fibroblasts cells (HTERT-BJ1). While Au-Tat particles are mainly located in the nucleus, only a few Au-tiopronin particles appear to be taken into the cell through membrane invaginations and accumulate in the cytoplasm in vacuoles or surrounding the mitochondria (Figure 3). Au-tiopronin nanoparticles are not detected in the nucleus in any case. This demonstrates the specificity of Au-Tat particles to be translocated and accumulate in the nucleus. This research has many potential applications in cell imaging, but also extends wider biomedical applications, such as gene therapy and drug delivery.

Peptide sequence RGD is also used for intracellular delivery of gold nanoparticles. But these peptides do not induce nuclear translocation. RGD receptors are overexpressed on the surface of human fibroblast cells. But it is relevant to mention that RGD peptide should be linked through an appropriate linking moiety to the AuNPs to avoid loss of RGD-mediated RME. AuNPs-tiopronin conjugates can be linked to RGD peptide through secondary linkers, such as ethylenediamine (EDA) and poly(ethylene glycol) bis (3-aminopropyl) terminated (PEG) [81]. As expected for a RME-sequenced peptide, RGD-EDA-tiopronin-AuNPs conjugates are internalized within the cytoplasm of the human fibroblast cells. However, in case of RGD-PEG-tiopronin-AuNPs conjugates no internalization is observed [70].

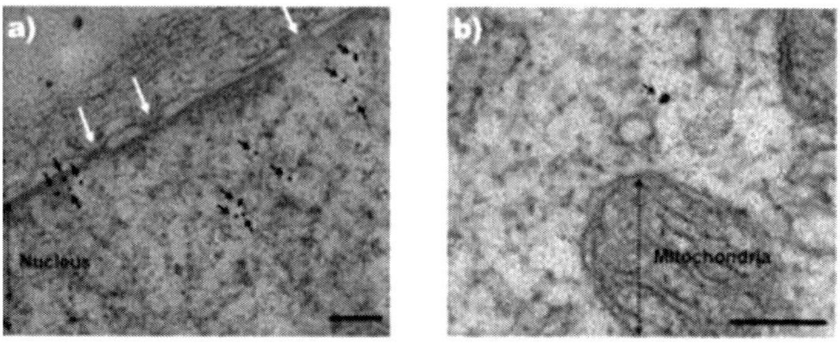

Reprinted with permission from ref (68). Copyright 2005, American Chemical Society.

Figure 3. Transmission electron micrographs of human fibroblasts incubated with Au@Tat (a) and Au@tiopronin (b) nanoparticles. The black dots indicated with arrows are nanoparticles, and white arrows show nuclear membrane pores. (scale bars) 50 nm).

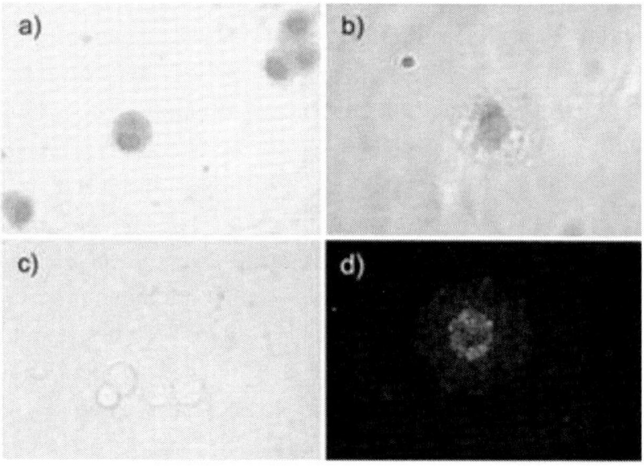

Reprinted with permission from ref (83). Copyright 2008, American Chemical Society.

Figure 4. Bright-field microscopic images of HeLa cell after being incubated with 0.04 nM 30 nm gold nanoparticle-peptide complexes for 24 h: (a) mixed peptides (CALNNR8/CALNN) 1:9), (b) pure CALNNR8, and (c) pure CALNN-functionalized gold nanoparticles. (d) Corresponding dark-field microscopic image of one cell in (a).

Though RGD-derived peptides are unable to enter nucleus, recently Tkachenko et al has discovered that modification of these peptides are capable of targeting the nucleus of HepG2 cells and HeLa cells, when conjugated with AuNPs. These modified peptides are composed of the integrin binding domain (RGD) and a stretch of six lysine residues [69]. Integrin binding domains are known to induce cellular uptake by both RME and phagocytosis. However, usually they do not interact with the nuclear pore complex. So, probably these conjugates enter the cell via the integrin binding domain and escape endosomes due to the presence of positively charged lysines, which are known to induce leakage of negatively charged vesicles and cells.

Recently, cell penetrating peptides have been used to deliver AuNPs into osteosarcoma (bone cancer) cells [82]. Here, we modified the peptides with cysteine terminus to attach with gold nanoparticles. Three types of peptides have been studied. Cell penetrating peptides (CPPs) and NLS attached gold nanoparticles efficiently enter inside cells. But exceptionally high uptake of conjugates by cells is observed when CPPs coupled with NLS. After 3h of incubation with the cells, most of the internalized peptide-nanoparticles conjugates are found clustered together inside membrane –bound cellular compartments that appear to be endosomes as observed by TEM.

Arginine-rich peptide ($CALNNR_8$)-functionalized gold nanoparticles have been prepared for the purpose of cellular membrane transportation where CALNN is used as a stabilizing agent [83]. The AuNPs-peptide complexes exhibit specific recognition properties for nucleus and endoplasmic reticulum targeting. The translocation ability of AuNPs-peptide complexes is dependent on the ratio of $CALNNR_8$ to CALNN on the particle surface and the nanoparticle diameter. Mixed peptide (CALNN and $CALNNR_8$)-functionalized AuNPs are capable of targeting the cell nucleus. However, AuNPs-peptide complexes synthesized by pure $CALNNR_8$ do not reach the nucleus (Figure 4). It indicates that not only the arginine part of the carrier peptides direct the internalized localization, but also the stabilizing peptide, CALNN, contributes to this process. Very recently, Pujals et al has shown that AuNPs can be shuttled into tumoral cells with an amphipathic proline-rich peptides, $(VRLPPP)_3$ which is also known as sweet arrow peptide (SAP) [84].

SMALL MOLECULE MEDIATED DELIVERY

AuNPs can be delivered by small molecule therapeutic agents into tumor. Upon accumulation at the tumor site, the attached small molecule facilitates a

RME mediated uptake of the AuNPs into the diseased cells. One example of such small molecule is folic acid (FA), a natural B-vitamine has been exploited to selectively target folate receptor (FR) expressing cells [85-89]. FR is a useful target for tumor-specific drug delivery because this is highly expressed on epithelial cancers of the ovary, brain, throat kidney, colon, lung, prostate, breast and mycloid cells etc. [86]. Cojugating FA and a therapeutic agent to metal nanoparticles is an important step for FR-mediated targeted drug delivery to cancer cells. AuNPs-folate conjugates show promising result in selective targeting of FR-positive tumors. These conjugates have been used in tumor imaging and photothermal therapy applications [90]. A poly(ethylene glycol) (PEG) construct with thioctic acid and folic acid coupled on opposite ends of the polymer chain has been synthesized with AuNPs for targeting the FR expressing tumor cells via RME. The resulting coated AuNPs are water soluble and are designed as (i) mPEG-thioctic acid-modified AuNPs (mPEG 2000 – T:AuNP) or (ii) the folate-PEG 1500- thioctic acid- modified AuNPs (F-PEG1500-T: AuNP). Internalization of these conjugates have been studied in KB cells (human epithelial carcinoma cell line that expresses the FR) and WI-38 cells (a normal embryonic human diploid cell line as a folate receptor deficient cells). Efficient internalization has been observed in case of tumor cells expressing the FRs. AuNPs-conjugates uptake is minimal in other types of cancer cells that lack FRs.

There are only few reports that describe the attachment of FA to metal-besed nanoparticles usig covalent interaction [90, 91]. Recently, a simple method of attaching FA to AuNPs has been demonstrated using noncovalent interaction [92]. Different PEG backbones has been used to tune the attachment process. Successful delivery has been achieved to different cancer cells (SKOV-3, OVCAR-5, OV-202, OV-167, OPM-1, RPMI and U266) with differential FR expression. Quantitation of the internalization of the nanoconjugates in different cell lines is determined by gold analysis with inductively Coupled plasma. Uptake of the nanoconjugates is correlated with FR expression. TEM imaging of treated cells reveals the endocytosis of the nanoconjugates. Dendromer-entrapped, folate functionalized AuNPs have also shown the potential for targeting and imaging cancer cells [93]. when KB cells treated with this conjugate selective uptake by the high FR-expressing cells has been observed. Methotrexate (MTX), an analogue of folic acid, when conjugated to AuNPs avoid tumor cell resistance which invariably developed upon repeated use of this flexible anticancer drug [94]. It is observed that MTX-AuNPs conjugates rapidly accumulated in LL2 (Lewis lung carcinoma) cells, inducing higher cytotoxic effects on the tumor compared with free MTX which shows no antitumor effects.

AuNPs conjugates of other chemotherapeutic agents have been reported to address various limitations of the unconjugated agents [95, 96]. A multifunctional vector consisting of AuNPs, TNFα, thiolated PEG and paclitaxel (PTX), a leading anticancer drug has been reported for targeted drug delivery to solid tumors [47]. The release of PTX from the vector is investigated in B16/F10 melanoma cancer cells. It is observed that the vector remains inactive unless treated with dithiothretol (DTT). It indicates that the vector is acting as a prodrug from which PTX must be released to show the desired anticancer effects. It is shown that PTX-PT-AuNPs-TNFα vector delivers 10-fold more TNFα and PTX to the tumor site when compared to unconjugated ones. Recently, Mukherjee et al has developed similar type of vector consisting of two components with different functions: an antiangiogenic molecule, VEGF antibody-2C3 (AbVF), and an anticancer drug, gemicitabine [97]. They have developed this '2 in 1' system onto a single AuNP core. AuNPs conjugates of fluorescent small molecule, such as coumarin, have been developed as cellular probes and delivery agents [98]. Attachement of PEG-functionalized AuNPs to coumarin through carbamate bond shows significant enhancement of emission intensity. AuNPs conjugates are efficiently internalized in MDA-MB-231 cells and localized in the perinuclear region as evidenced through intracellular particle tracking. Very recently, Kahalalide F, an antitumoral drug has been conjugated with gold nanoparticles [99].

The conjugation of AuNPs to an antitumoral drug could play two parallel roles: to enhance the delivery of a drug to the cell by presenting numerous peptide copies anchored to the nanoparticle surface and to help in the delivery of the nanoparticle to the pathological cell favoring the local aggregation of nanoparticles in tumor cells by increasing absortion in the NIR where tissues present low NIR region radiation absortion, after irradiation. Kahalalide F analogue has been conjugated with AuNPs to enhance the delivery of Kahalalide F to tumoral cells by increasing its cytotoxic effect. Cellular uptake of conjugates has been studied into HeLa cells. Two different sizes of AuNPs , i.e., 20 and 40 nm (AuNP-20, AuNP-40, respectively) have been made to determine the influence of nanoparticle size on the penetration and cytotoxicity of the conjugates. As expected, the accumulation of AuNP-40 is higher than AuNP-20. AuNP-40 conjugates also show higher cytotoxicity with respect to the free peptides. This enhancement could be due to better cell uptake of the former conjugate.

Liposomes can be used as a 'Trojan Horse' to deliver small (1.4 nm) AuNPs into tumor cells [100]. It is revealed that the liposomal approach provides a thousand-fold enhancement in the cellular uptake of the small AuNPs. Real-time intracellular tracking of the AuNP-liposomes reveals an average speed of 12.48±3.12 um/hr for their intracellular transport.

CONCLUSION AND FUTURE OUTLOOK

In this chapter, I have focused on different vehicles for internalization of gold nanoparticles inside cells and applications of internalized particles in nanomedicine. Several ligands such as peptides, proteins and small molecules have been successfully used for specific delivery of AuNPs. It is now possible to control the size of gold nanoparicles at nanometer resolution. Additionally, the biocompatibility and photo-optical distinctiveness of gold nanoparticles are now proven to be powerful in diagnostic applications.

However, a very important issue in the successful implementation of the promised applications of AuNPs is toxicity. There are many reports available on non-toxicity of AuNPs. AuNPs emerge much less toxic than other types of nanoparticles. Some of the ligands used for AuNPs-conjugation are moderately toxic. Many biocompatible ligands including thiolated PEGs are non-toxic. Another issue is the poor reproducibility and wide range of size distribution of nanoparticle product synthesized from wet chemical methods. Improved synthetic methodologies that can lead to true monodispersity and satisfactory reproducibility of nanoparticles with well-controlled shape remain to be developed.

Recent reports show that AuNPs –based therapeutic agents could overcome the barriers presented by the human immune systems to achieve delivery at diseased sites without effecting healthy tissues. In conclusion, AuNPs are easily bioconjugable and very promising for imaging, diagnostics and therapy biomedical applications for cancer and a number of other diseases for human.

REFERENCES

[1] Finley, RS. Overview of targeted therapies for cancer. *Am. J. Health Syst. Pharm.*, 2003, 60, S4-10.

[2] Portlock, JL; Calos, MP. Site-specific genomic strategies foe gene therapy. *Curr. Opin. Mol. Ther.*, 2003, 5, 376-382.

[3] Abou-Jawde, R; Choueiri, T; Alemany, C; Mekhail, T. An overview of targeted treatments in cancer. *Clin. Ther.*, 2003, 25, 2121-2137.

[4] Weissig, V. Mitochonrial-targeted drug and DNA delivery. *Crit. Rev. Ther. Drug Carrier Syst.*, 2003, 20, 1-62.

[5] Russel-Jones, GJ. The potential use of receptor-mediated endocytosis for oral drug delivery. *Adv. Drug Delivery Rev.*, 2001, 46, 59-73.

[6] Hayat, MA. (Ed). *Colloidal gold: Principles, Methods, and Apllications*, Academic press, San Diego, 1989.

[7] Schultz, DA. Plasmon resonant particles for biological detection. *Curr. Opin. Biotech.*, 2003, 14, 13-22.

[8] Daniel, MC; Astruc D. Gold nanoparticles: assembly, supramolecular chemistry, quatum-size-related properties, and applications toward biology, catalysis, and nanotechnology. *Chem. Rev.*, 2004, 104, 293-346.

[9] Bulmus, V. Biomembrane-active molecular switches as tools for intracellular drug delivery. *Aus. J. Chem.*, 2005, 58, 411-422.

[10] Deshayes, S; Morris, MC; Divita, G; Heitz, F. Interactions of primary amphiphatic cell penetrating peptides with model membrane: consequences on the mechanisms of intracellular delivery of therapeutics. *Curr. Pharm.*, 2005, 11, 3629-3638.

[11] Panyam, J; Labhasetwar, V. Targeting intracellular targets. *Curr. Drug Delivery*., 2004, 1, 235-247.

[12] Chavanpatil, MD; Khdair, A; Panyam, J. Nanoparticles for cellular drug delivery: mechanisms and factors influencing delivery. *Journal of Nanoscience and Nanotechnology*., 2006, 6(9-10), 2651-63.

[13] Nahar, M; Dutta, T; Murugesan, S; Asthana, A; Mishra, D; Rajkumar, V. et al. Functional polymeric nanoparticles: an efficient and promising tool for active delivery of bioactives. *Critical Reviews in Therapeutic Drug Carrier Systems*., 2006, 23(4), 259-318.

[14] Nimmagadda, A; Thurston, K; Nollert, MU; McFetridge, PS. Chemical modification of SWNT alters in vitro cell–SWNT interactions. *Journal of Biomedical Materials Research* A., 2006, 76(3), 614-25.

[15] Reddy, GR; Bhojani, MS; McConville, P; Moody, J; Moffat, B. A; Hall, D. E, et al. Vascular targeted nanoparticles for imaging and treatment of brain tumors. *Clinical Cancer Research*., 2006, 12(22), 6677-86.

[16] Moulari, B; Pertuit, D; Pellequer, Y; Lamprecht, A. The targeting of surface modified silica nanoparticles to inflammed tissue in experimental colitis. *Biomaterials*., 2008, 29(34), 4554-60.

[17] Shaw III, FC. Gold –based therapeutic agents. *Chem. Rev.*, 1999, 99, 2589-2600.

[18] Frens, G. Controlled nucleation for the regulation of the particle size in monodisperse gold suspensions. *Nature: Phys. Sci.*, 1973, 241, 20-22.

[19] Brust, M; Walker, M; Bethell, D; Schif frin, DJ; Whyman, R. Synthesis of thiol-derivatized gold nanoparticles in a 2-phase liquid–liquid system. *J. Chem. Soc.-Chem. Commun.*, 1994, 801-802.

[20] Esumi, K; Suzuki, A; Yamahira, A; Torigoe, K. Role of poly(amidoamine) dendrimers for preparing nanoparticles of gold, platinum, and silver. *Langmuir.*, 2000, 16, 2604-2608.

[21] Brown, KR; Natan, MJ. Hydroxylamine seeding of colloidal Au nanoparticles in solution and on surface. *Langmuir.*, 1998, 14, 726-728.

[22] Jana, NR; Geraheart, L; Murfy, CJ. Seed-mediated growth approach for shape-controlled synthesis of spheroidal and rodlike gold nanoparticles using a surfactant template. *J. Adv. Mater.*, 2001, 13, 1389-1393.

[23] Tsuji, M; Miyamae, N; Hashimoto, M; Nishio,M; Hikino, S; Ishigami, N; Tanaka, I. Shape and size controlled synthesis of gold nanocrystals using oxidative etching by $AuCl^{-4}$ and Cl^{-} anion in microwave-polyol process. *Coll. Surf. A.*, 2007, 302, 587-598.

[24] Dahl, JA; Maddux, BLS; Hutchison, JE. Toward greener nanosynthesis. *Chem. Rev.*, 2007, 107, 2228-2269.

[25] Li, ZH; Liu, ZM; Zhang, JL; Han, BX; Du, JM; Gao, YN; Jiang, T. Synthesis of single-crystal gold nanosheets of large size in ionic liquids. *J. Phys. Chem. B* 2005, 109, 14445-14448.

[26] Jin, Y; Wang, PJ; Yin, DH; Liu, JF; Qin, LS; Yu, NY; Xie, GY ;Li, BM. Gold nanoparticles prepared by sonochemical method in thiol-functionalized ionic liquid. *Coll. Surf. A.*, 2007, 302, 366-370.

[27] Li, CC; Cai, WP; Li, Y; Hu, JL; Liu, PS. Ultrasonically induced Au nanoprisms and their size manipulation based on aging. *J. Phys. Chem. B.*, 2006, 110, 1546-1552.

[28] Pol, VG; Wildermuth, G; Felsche, J; Gadenken, A; Calderon-Moreno, J. Sonochemical deposition of Au nanoparticles on titania and the significant decrease in the melting point of gold. *J. Nanosci. Nanotechnol.*, 2005, 5, 975-979.

[29] Park, JE; Momma, T; Osaka, T. Spectroelectrochemical phenomena on surface plasmon resonance of Au nanoparticles immobilized on transparent electrode. *Electrochimica Acta.*, 2007, 52, 5914-5923.

[30] Abyaneh, MK; Paramanik, D; Varma, S; Gosavi, SW; Kulkarni, SK. Formation of gold nanoparticles in polymethylmethacrylate by UV irradiation. *J. Phys. D-Appl. Phys.*, 2007, 40, 3771-3779.
[31] Luo, XZ. and Imae, T. Shape-controlled synthesis of gold nanoparticles under UV irradiation in the presence of poly (ethylene glycol). *Curr. Nanosci.*, 2007, 3, 195-198.
[32] Chen, Y; Cho, J; Young, A; Taton, TA. Enhanced stability and bioconjugation of photo-cross-linked polystyrene-shell, Au-core nanoparticles. *Langmuir.*, 2007, 23, 7491-7497.
[33] Tilaki, RM; Zad, IA; Mahdavi, SM. The effect of liquid environment on size and aggregation of gold nanoparticles prepared by pulsed laser ablation. *J. Nanopart. Res.*, 2007, 9, 853-860.
[34] Giorgetti, E; Giusti, A; Laza, SC; Marsili, P; Giammanco, F. Production of colloidal gold nanoparticles by picosecond laser ablation in liquids. *Phys. Status Solidi A.*, 2007, 204, 1693-1698.
[35] Guo, S; Shi, F; Gu, YL; Yang, J; Deng, YQ. Size-controllable synthesis of gold nanoparticles via carbonylation and reduction of hydrochloroauric acid with CO and H2O in ionic liquids. *Chem. Lett.*, 2005, 34, 830-831.
[36] Nakamoto, M; Kashiwagi, Y; Yamamoto, M. Synthesis and size regulation of gold nanoparticles by controlled thermolysis of ammonium gold(I) thiolate in the absence or presence of amines. *Inorg. Chim. Acta.*, 2005, 358, 4229-4236.
[37] Mandal, TK; Fleming, MS; Walt, DR. Preparation of polymer coated gold nanoparticles by surface-confined living radical polymerization at ambient temperature. *Nano Lett.*, 2002, 2, 3-7.
[38] Kell, AJ; Alizadeh, A; Yang, L; Workentin, MS. Monolayerprotected gold nanoparticle coalescence induced by photogenerated radicals. *Langmuir.*, 2005, 21, 9741-9746.
[39] Gachard, E; Remita, H; Khatouri, J; Keita, B; Nadjo, L; Belloni, J. Radiation-induced and chemical formation of gold clusters. *New J. Chem.*, 1998, 22, 1257-1265.
[40] Warren, SC; Jackson, AC; Cater-Cyker, ZD; DiSalvo, FJ; Wiesner, U. Nanoparticle synthesis via the photochemical polythiol process. *J. Am. Chem. Soc.*, 2007, 129, 10072-10073.
[41] Mandal, D; Bolander, ME; Mukhopadhyay, D; Sarkar, G; Mukherjee, P. The use of microorganisms for the formation of metal nanoparticles and their application. *Appl. Microbiol. Biotechnol.*, 2006, 69, 485-492.

[42] Mukherjee, P; Ahmad, A; Mandal, D; Senapati, S; Sainkar, SR; Khan,MI; Ramani, R; Parischa, R; Ajayakumar, PV; Alam, M; Sastry, M; Kumar, R. Bioreduction of $AuCl_4^-$ Ions by the Fungus, Verticillium sp. and Surface Trapping of the Gold Nanoparticles Formed. *Angew. Chem. Int. Ed.*, 2001, 40, 3585-3588.

[43] Mukherjee, P; Senapati, S; Mandal, D; Ahmad, A; Khan,MI; Sastry, M; Kumar, R. Extracellular synthesis of gold nanoparticles by the fungus *Fusarium Oxysporum. Chem. BioChem.*, 2002, 5, 461-463.

[44] Woodle, MC; Engbers, CM; Zalipsky S. New amphipatic polymer-lipid conjugates forming long-circulating reticuloendothelial system-evading liposomes. *Bioconjug. Chem.*, 1994, 5, 493-496.

[45] Raynal, I; Prigent, P; Peyramaure, S; (Eds.), Macrophage endocytosis of superparamagnetic iron oxide nanoparticles: mechanisms and comparision of ferumoxides and ferumoxtran-10. *Invest. Radiol.*, 2004, 39, 56-63.

[46] Roger, WJ; Basu, P. Factors regulating macrophage endocytosis of nanoparticles: implications for targeted magnetic resonance plaque imaging. *Atherosclerosis.*, 2005, 178, 67-73.

[47] Paciotti, GF; Kingston, DGI; Tamarkin, L. Colloidal gold nanoparticles: a novel nanoparticle platform for developing multifunctional tumor-targeted drug delivery vectors. *Drug Dev. Res.*, 2006, 67, 47-54.

[48] Paciotti, GF; Myer, L; Weireich, D; (Eds.), Colloidal gold: A novel nanoparticle vector for tumor directed drug delivery. *Drug Deliv.*, 2004, 11, 169-183.

[49] Visaria, RK; Griffin, RJ; Williams, BW; Ebbini, ES; Paciotti, GF; Song, CW; Bischof, JC. Enhancement of tumor thermal therapy using gold nanoparticle-assisted tumor necrosis factor-α delivery. *Mol. Cancer Ther.*, 2006, 5, 1014-1020.

[50] Srinivasan, JM; Fajardo, LF; Hahn, GM. Mechanism of antitumor activity of tumor necrosis factor alpha with hyperthermia in a tumor necrosis factor alpha-resistant tumor. *J. Natl. Cancer Inst.*, 1990, 82, 1904-1910.

[51] Umeno, H; Watanabe, H; Yamauchi, N; et al. Enhancement of blood stasis and vascular permeability in Meth-A tumors by administration of hyperthermia in combination with tumor necrosis factor. *Jpn. J. Cancer Res.*, 1994, 85, 325-330.

[52] Arteaga, CL. The epidermal growth factor receptor: from mutant oncogene in nonhuman cancers to therapeutic target in human neoplasia. *J. Clin. Oncol.*, 2001, 19, 32s-40s.

[53] Xiong, HQ; Abbruzzese, IL. Epidermal growth factor receptor-targeted therapy for pancreatic cancer. *Semin. Oncol.*, 2002, 29, 31-37.

[54] Paez, JP; Janne, PA; Lee, JC; et al. EGFR mutations in lung cancer: correlation with clinical response to gefitinib therapy. *Science.*, 2004, 304, 1497-1500.

[55] Ahmed, SM; Salgia, R. Epidermal growth factor receptor mutations and susceptibility to targeted therapy in lung cancer. *Respirology.*, 2006, 11, 687-692.

[56] Prudkin, L; Wistuba II. Epidermal growth factor receptor abonormalities in lung cancer. Pathogenetic and clinical implications. *Ann. Diagn. Pathol.*, 2006, 10, 306-315.

[57] Cohenuram, M; Saif, MW. Epidermal growth factor receptor inhibition strategies in pancreatic cancer: past, present and the future. *J. Pancreas.*, 2007, 8, 4-15.

[58] El-Sayed, IH; Huang, XH; El-Sayed, MA. Surface Plasmon resonance scattering and absorption of anti-EGFR antibody conjugated gold nanoparticles in cancer diagnostics: Applications in oral cancer. *Nano Lett.*, 2005, 5, 829-834.

[59] Huang, XH; El-Sayed, IH; Qian, W; El-Sayed, MA. Cancer cell imaging and photothermal therapy in the near-infrared region by using gold nanorods. *J. Am. Chem. Soc.*, 2006, 128, 2115-2120.

[60] El-Sayed, IH; Huang, XH; El-Sayed, MA. Selective laser photo-thermal therapy of epithelial carcinoma using anti-EGFR antibody conjugated gold nanoparticles. *Cancer Lett.*, 2006, 239, 129-135.

[61] Qian, X; Peng, XH; Ansari, DO; (Eds,), *In vivo* tumor targeting and spectroscopic detection with surface-enhanced Raman nanoparticle tags. *Nat. Biotechnol.*, 2008, 26,83–90.

[62] Qian, ZM; Li, H; Sun, H; Ho, K. Targeted drug delivery via the transferring receptor-mediated endocytosis pathway. *Pharmacol. Rev.*, 2002, 54, 561-587.

[63] Yang, PH; Sun, X; Chiu, JF. (Eds.), Transferrin-mediated gold nanoparticle cellular uptake. *Bioconjug Chem.*, 2005, 16, 494-496.

[64] Li , JL; Wang, L; Liu, XY; Zhang, ZP; Guo, HC; Liu, WM; Tang, SH. In-vitro cancer cell imaging and therapy using transferring-conjgated gold nanoparticles. *Cancer Lett.*, 2009, 274, 319-326.

[65] Tkachenko, AG; Xie, H; Coleman, D. (Eds.), Multifunctional gold nanoparticle-peptide complexes for nuclear targeting. *J. Am. Chem. Soc.*, 2003, 125, 4700-4701.

[66] Kalderon, D; Richardson, WD; Markham, AF. et al. Sequence requirements for nuclear location of simian virus 40 large-T antigen. *Nature (London)* ., 1984, 311, 33-38.

[67] Oyelere, AK; Chen, PC; Huang, X; et al. Pepetide-conjugated gold nanorods for nuclear targeting. *Bioconjug Chem.*, 2007, 18, 1490-1497.

[68] de la Furente, JM; Berry, CC. Tat peptide as an effi cient molecule to translocate gold nanoparticles into the cell nucleus. *Bioconjug Chem.*, 2005, 16, 1176-1180.

[69] Tkachenko, AG; Xie, H; Coleman, D; (Eds.), Cellular trajectories of peptide-modifi ed gold particle complexes: comparison of nuclear localization signals and peptide transduction domains. *Bioconjug Chem.*, 2004, 15,482-490.

[70] de la Furente, JM; Berry, CC; Riehle, MO; (Eds.), Nanoparticle targeting at cells. *Langmuir.*, 2006, 22, 3286-93.

[71] Liu, Y; Shipton, MK; Kaufman, ED; (Eds.), Synthesis, stability, and cellular internalization of gold nanoparticles containing mixed peptidepoly(ethylene glycol) monolayers. *Anal. Chem.*, 2007, 79, 2221-2229.

[72] Feldherr, MC; Akin, DJ. The permeability of the nuclear envelope in dividing and nondividing cell cultures. *J. Cell Biol.*, 1990, 111,1-8.

[73] Feldherr, CM; Lanford, RE; Akin, D. Signal-mediated nuclear transport in simian virus 40-transformed cells is regulated by large tumor antigen. *Proc. Natl. Acad. Sci. USA.*, 1992, 89, 11002-11005.

[74] Butel, JS. Jarvis, DL. The plasma membrane associated form of SV-40 large tumor antigen: biochemical and biological properties. *Biochim. Biophys. Acta.*, 1986, 865, 171-195.

[75] De Pamphilis, ML; Bradly, MK. Replication of SV40 and polyoma virus chromosomes. In: Saltzman, NP., Ed. *The Papovaviridae*. New York: Plenum; 1986, Vol 1, 99-246.

[76] Robbins, J; Dilworth, SM; Lasky, RA; Dingwall, C. Two independent basic domains in nucleoplasmic nuclear targeting sequence identification of a class of bipartite nuclear targeting sequence. *Cell.*, 1991, 64, 615-623.

[77] Feldherr, CM; Kallenbach, E; Schultz, N. Movement of a karyophilic protein through the nuclear pores of oocytes. *J. Cell. Biol.*, 1984, 99, 2216-2222.

[78] Ryan, JA; Overton, KW; Speight, ME; (Eds.), 2007. Cellular uptake of gold nanoparticles passivated with BSA-SV40 large t antigen conjugates. *Anal. Chem.*, 2007, 79, 9150-9159.

[79] Lewin, M; Carlesso, N; Tung, CH, et al. Tat peptide-derivatized magnetic nanoparticles allow in vivo tracking and recovery of progenitor cells. *Nat. Biotechnol.*, 2000, 18, 410-414.

[80] Ford, KG; Souberbielle, BE; Darling, D; (Eds.), Protein transduction: an alternative to genetic intervention? *Gene Ther.*, 2001, 8, 1-4.

[81] Mrksich, M; Whitesides, GM. Using self-assembled monolayers to understand the interactions of man-made surfaces with proteins and cells. *Annu. Rev. Biophys. Biomol. Struct.*, 1996, 25, 55-78.

[82] Mandal, D; Maran, A; Yaszemski, MJ; Bolander, ME; Sarkar, G. Cellular uptake of gold nanoparticles directly cross-linked with carrier peptides by osteosarcoma cells. *J. Mater Sci. Mater Med.*, 2009, 20(1), 347-50.

[83] Sun, L; Liu, D; Wang, Z. Functional gold nanoparticle-peptide complexes as cell-targeting agents. *Langmuir.*, 2008, 24, 10293-10297.

[84] Pujals, S; Bastus, NG; Pereiro, E; Lopez-Iglesias, C; Puntes, VF; Kogan, MJ; Giralt, E. Shuttling gold nanoparticles into tumoral cells with an amphipathic proline-rich peptide. *Chem. BioChem.*, 2009, 10,1025-1031.

[85] Sudimack, J; Lee, RJ. Targeted drug delivery via the folate receptor. *Adv. Drug Deliv. Rev.*, 2000, 41, 147-162.

[86] Lu, Y; Low, PS. Folate-mediated delivery of macromolecular anticancer therapeutic agents. *Adv. Drug Deliv. Rev.*, 2002, 54, 675-693.

[87] Lu, Y; Sega, E; Leamon, CP; (Eds.), Folate receptor-targeted immunotherapy of cancer: mechanism and therapeutic potential. *Adv. Drug Deliv. Rev.*, 2004, 56, 1161-1176.

[88] Roy, EJ; Gawlick, U; Orr, BA; (Eds.), Folate-mediated targeting of T cells to tumors. *Adv. Drug Deliv. Rev.*, 2004, 56, 1219-1231.

[89] Stevens, PJ; Sekido, M; Lee, RJ. A folate receptor-targeted lipid nanoparticle formulation for a lipophilic paclitaxel prodrug. *Pharm. Res.*, 2004, 21, 2153-2157.

[90] Dixit, V; Van den Bossche, J; Sherman, DM; (Eds.), Synthesis and grafting of thioctic acid-PEG-folate conjugates onto Au nanoparticles for selective targeting of folate receptor-positive tumor cells *Bioconjug Chem.*, 2006, 17, 603-609.

[91] Kohler, N; Fryx ell, GE; Zhang, MQ. A bifunctional poly(ethylene glycol) silane immobilized on metallic oxide-based nanoparticles for conjugation with cell targeting agents. *J. Am. Chem. Soc.*, 2004, 126, 7206.

[92] Bhattacharya, R; Patra, CR; Earl, A; (Eds.), Attaching folic acid on gold nanoparticles using noncovalent interaction via different polyethylene

glycol backbones and targeting of cancer cells. *Nanomedicine.*, 2007, 3, 224-238.

[93] Shi, X; Wang, S; MeshinchI, S; (Eds.), Dendrimer-entrapped gold nanoparticles as a platform for cancer-cell targeting and imaging. *Small.*, 2007, 3,1245-1252.

[94] Chen, Y; Tsai, C; Huang, P; (Eds.), Methotrexate conjugated to gold nanoparticles inhibits tumor growth in a syngeneic lung tumor model. *Mol. Pharm.*, 2007, 4, 713-722.

[95] Ganesh, T. Improved biochemical strategies for targeted delivery of taxoids. *Bioorg. Med. Chem.*, 2007, 15, 3597-3623.

[96] Vijayaraghavalu, S; Raghavan, D; Labhasetwar, V. Nanoparticles for delivery of chemotherapeutic agents to tumors. *Curr. Opin. Investig. Drugs.*, 2007, 8, 477-484.

[97] Mukherjee, P; Bhattacharya, R; Mukhopadhyay, D. Gold nanoparticles bearing functional anti-cancer drug and anti-angiogenic agent: A "2 in 1" system with potential application in cancer therapeutics. *J. Biomed. Nanotech.*, 2005, 1, 224-228.

[98] Shenoy, D; Fu, W; Li, J., et al. Surface functionalization of gold nanoparticles using hetero-bifunctional poly(ethylene glycol) spacer for intracellular tracking and delivery. *Int. J. Nanomedicine.*, 2006, 1, 51-57.

[99] Hosta, L; Pla-Roca, M; Arbiol, J; Lopez-Iglesias, C; Samitier, J; Cruz, LJ; Kogan, MJ; Albericio, F. Conjugation of Kahalalide F with gold nanoparticles to enhance in vitro antitumoral activity. *Bioconjug. Chem.*, 2009, 20, 138-146.

[100] Chithrani, DB; Dunne, M; Stewart, J; Allen, C; Jaffray, DA. Cellular uptake and transport of gold nanoparticles incorporated in a liposomal carrier. *Nanomedicine.*, 2009 (In press).

In: Gold Nanoparticles ISBN 978-1-62257-927-3
Editors: A. Jarnagin and L. Halshauser

Chapter 7

CHEMICAL SYNTHESIS OF GOLD NANOPARTICLES USING FLUOROQUINOLONE AND THEIR CHARACTERIZATIONS

S. Ashok Kumar[1,*], Sea-Fue Wang[1] and Yu-Tsern Chang[2]

[1]Department of Materials and Mineral Resources Engineering, National Taipei University of Technology, Taipei, Taiwan

[2]Department of Chemical and Materials Engineering, Nanya Institute of Technology, Jhongli, Taiwan

ABSTRACT

We report synthesis of water-soluble gold nanoparticles (Au-NPs), having an average diameter of ca. 20 nm, by reducing AuCl3 using ciprofloxacin (CF) as a reducing/stabilizing agent. The synthesized Au-NPs have been characterized by scanning electron microscopy (SEM), EDS, TEM, UV-visible spectroscopy, X-ray diffraction and cyclic voltammetry. TEM and SEM combined with EDS confirmed that spherical shaped Au-NPs were formed. Uv-vis spectra of Au-NPs show two absorption bands corresponding to the capping agent ciprofloxacin and surface plasmon absorption bands at 274 nm and 535 nm,

* E-mail address: sakumar80@gmail.com.

respectively. The Au-NPs are used to modify a glassy carbon electrode (GCE) and it's electrochemical and electrocatalytic properties are investigated. The Au-NPs show excellent electrocatalytic activity toward the oxidation of methanol at +0.33 V in alkaline solution, which was not observed on unmodified GCE. Further, Au-NPs modified electrode showed excellent electrocatalytic properties for reduction of oxygen at lower potential. Here, CF was used as a reducing agent to synthesis of Au-NPs. The synthesized Au-NPs are highly stable, can be stored for more than three months in air at room temperatures.

1. INTRODUCTION

Synthesis of stable gold nanoparticles (Au-NPs) and their applications has been the subject of great interest, due to their remarkable physical and chemical properties. Au-NPs have uniform structures with promising stability and their size-related electronic, magnetic, and optical properties [1, 2] make them promising in important research areas such as catalysis [3], biosensors [4], drug delivery [5], energy-related applications [6], biological applications [7] etc. Controlling the size of nanoparticles (NPs) has always been one of the challenges in colloidal science. Changing the size of NPs can result in modulation of their physical and chemical properties. Since the discovery of various reducing agents for the gold compounds to form the gold nanoparticles, like sodium citrate, sodium borohydride, phosphorus, alcohols, and tannic acid/citrate mixtures, the synthesis and applications of Au-NPs of different sizes have flourished [8]. Recently, gold nanoparticles capped with novel zwitterionic disulfide ligands was reported which showed remarkable stability in saline media [9]. Cetyltrimethylammonium bromide/silver bromide complex as the capping agent of gold nanorods [10] and preparation of gold nanoparticles using ascorbic acid as reducing agent in reverse micelles [11] were reported. Au-NPs in the range of 5 to 6 nm were synthesized using sodium borohydride in the presence of newly synthesized mono-6-deoxy-6-pyridinium-β-cyclodextrin chloride [12]. In all those methods, different reducing and capping agents were used together to prepare stable Au-NPs. In few studies, some reagents were used as a reducing agent as well as stabilizing agent for Au-NPs.

Gold nanoflowers were obtained by a one-pot synthesis using N-2-hydroxyethylpiperazine-N-2-ethanesulphonic acid as a reducing/stabilizing agent and their electrocatalytic effect towards oxidation of methanol was studied [13]. Zhang et al reported synthesis of various gold nanostructures on glassy carbon electrodes in a low concentration of $HAuCl_4$ solution (5 mM) and enzyme-free sensor developed for the detection of glucose in pH 7.4 phosphate buffer solutions [14]. Tom et al. reported the antibacterial drug ciprofloxacin (CF) to protect gold nanoparticles of two different mean diameters, 4 and 20 nm. In their study, the trisodium citrate acted as capping agent and sodiumborohydride as the reducing agent. Further, they have investigated the nature of binding between Au-NPs and CF by several analytical techniques and proposed that the nitrogen atom of the NH moiety of piperazine group binds on the gold surface, as revealed by voltammetric and spectroscopic studies [15].

In this chapter, we discuss about a simple method to prepare stable, and water-soluble Au-NPs dispersion by ciprofloxacin [1-cyclopropyl-6-fluoro-1,4-dihydro-4-oxo-7-piperazinylquinolone-3-carboxylic acid] without using another reagent. Ciprofloxacin is belongs to quinolones which is a large and constantly expanding group of synthetic antibacterial agents [16–18], and ciprofloxacin is one of the most popular members of this family. Here, we examined a templateless, surfactantless chemical approach to the preparation of Au-NPs in a $HAuCl_4$ solution using lower concentration of ciprofloxacin (2 mM). We can easily control the morphologies of the final Au-NPs by simply changing the solution pH of precursors. The experimental results showed that the spherical Au-NPs obtained and exhibited high electrochemical activities for the oxidation of methanol, and reduction of oxygen.

2. Experimental

2.1. Reagents

$HAuCl_4.3H_2O$ and ciprofloxacin were purchased from Sigma-Aldrich. Potassium ferricyanide, potassium ferrocyanide, and disodium hydrogen phosphate were received from J. T. Baker. Methanol (99.9%) was purchased from ECHO, Taiwan. All other reagents used were analytical grade.

2.2. Instruments

The morphological features and element compositions were measured by field emission scanning electron microscope (SEM) and energy-dispersive X-ray (EDX) (HITACHI S-4700). Absorption spectra were recorded using a Uv–

vis spectrophotometer (PerkinElmer Lambda 900). Electrochemical experiments were performed with a CH Instruments (Chi611c, USA). All electrochemical experiments were carried out with a conventional three-electrode system. Glassy carbon electrode (GCE) was used as a working electrode and indium tin oxide (ITO) glass was used for the preparation of dry films. Platinum wire and Ag/AgCl (3 M KCl) were used as counter electrode and reference electrode, respectively. Electrolyte solution was purged with high-purity argon gas for 10 min prior to each electrochemical experiment. X-ray diffraction (XRD, Rigaku DMX-2200) with Cu-Kα radiation was conducted to reveal the nanostructures of the Au-NPs.

2.3. Preparation of Au-NPs

An aqueous solution of tetrachloroauric acid ($HAuCl_4.3H_2O$, 2 mM) was mixed with the solution of ciprofloxacin (dissolved in 0.1 M HCl) (2 mM) and the mixture was stirred using a magnetic bar at 1000 rpm, and the bath temperature was maintained at 60°C. After the addition of $HAuCl_4.3H_2O$ into CF solution, yellow color solution was formed.

The pH of the resulting mixture was adjusted from 2 to 8.0 using 0.5 M NaOH with constant stirring. In high pH, yellow solution was completely disappeared. After few min, wine red solution was obtained, and stored at room temperature. The obtained Au-NPs solution was centrifuged and the precipitate was washed five times with 0.1 HCl and, washing is repeated with de-ionized water for five more times to remove any unadsorbed or unreacted reagents. The washed precipitate can be dispersed easily in de-ionized water by sonication. The obtained Au-NPs dispersion was remain stable for more than three months at room temperatures.

3. Results and Discussions

As sown in Figure 1, $HAuCl_4$ solution (b) (2 mL, 2 mM) was mixed with the solution of CF (a) (2 mL, 2 mM) with continuous stirring at 1000 rpm using a magnetic bar. After mixing, immediately a yellow color solution was obtained as shown in Figure 1 (c). The color of the mixture was remains stable, however, yellow color precipitate was settled down when stopping the stirring process.

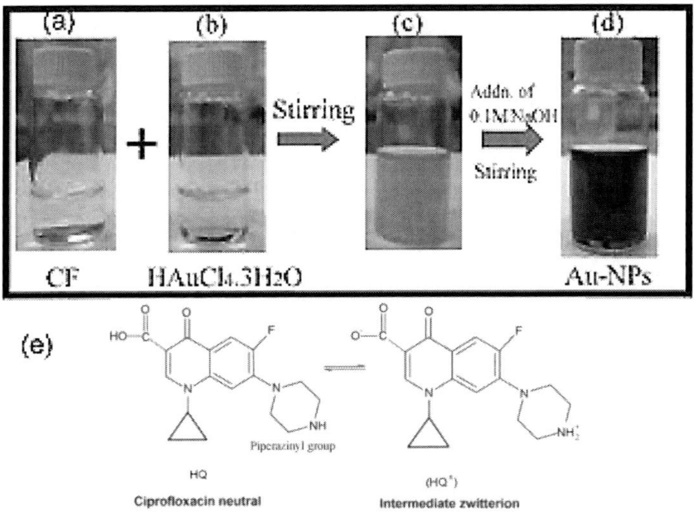

Figure 1. Representative experimental procedure for synthesis of Au-NPs by CF. (a) CF solution (2 mM, 2mL), (b) $HAuCl_4.3H_2O$ (2 mM, 2 mL), (c) mixture of CF + $HAuCl_4.3H_2O$ and (d) as-synthesizes Au-NPs. (e) Chemical formula of CF.

The pH of the $HAuCl_4$ and CF mixture was around 2. Consequently, pH of the mixture was adjusted by addition of base (0.5 M NaOH) with constant stirring. During this process, yellow color was disappeared and, we stopped addition of NaOH. After few min, wine-red color was started to appear and became intense with the time. A dark wine-red color solution was obtained after 5 min of stirring (Figure 1d).

Figure 2 shows the Uv-vis spectra of CF solution (curve c), $HAuCl_4$ solution (curve b) and as-prepared Au-NPs (curve a) colloidal solution. The absorption spectra of CF show three distinct absorption bands at 276, 314, and 327 nm (curve c). The absorption maximum at 276 nm corresponds to the π-π^* transition of the fluorobenzene moiety, and other two correspond to n-π^* as well as π-π^* transitions of the quinolone ring (Figure 1e) [15, 19, 20]. $HAuCl_4$ did not show any absorption peaks in the range from 200 to 400 nm (curve b). As expected, the absorption spectrum of synthesized Au-NPs solution shows four distinct bands at 270, 321, 333 and 527 nm (curve a). After generation of Au-NPs in CF solution, the absorption peak at 276 nm is shifted to shorter wavelength and other two bands also shifted to longer wave regions. In addition, a new band was appeared at 527 nm which is ascribed to the surface plasmon absorption of spherical Au-NPs. These observations were corroborating the formation of Au (0) nanoparticles [21].

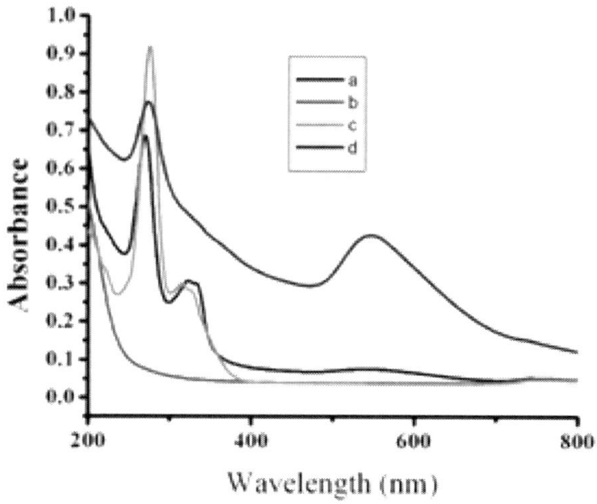

Figure 2. Uv-vis spectra of (a) Au-NPs/CF, (b) $HAuCl_4.3H_2O$, (c) CF solution and (d) separated Au-NPs re-dispersed in water.

To remove the un-reacted molecules, the synthesized Au-NPs were centrifuged; resulting CF-capped Au-NPs were washed with de-ionized water and 0.1 M HCl. The absorption spectrum of the separated Au-NPs was measured with the dilution in water, and it was found that two strong absorption bands at 274 and 527 nm (Figure 2 curve d). It was vouched that Au-NPs were prepared and the synthesized particles were covered by CF molecules. In the control experiment, synthesis of Au-NPs procedure was repeated without using CF solution; it was not observed any color change in $HAuCl_4$ solution. By this experiment, it was confirmed that CF molecules were worked as a reducing agent as well as capping agent for Au-NPs. The CF-protected Au-NPs can be easily dispersed in distilled water by sonication. A review by Turel discusses about the interactions of metal ions with quinolone antibacterial agents and the possible explanations were suggested [22]. Ciprofloxacin with the piperazinyl group in the 7-position contain two relevant ionizable functional groups. The protolytic equilibria of fluoroquinolone analogues are expressed in Figure 1e [22].

CF molecules can exist in four possible forms: an acidic cation H_2Q^+, a neutral nonionized species HQ, an intermediate zwitterion $HQ^{\pm}$ and a basic anion Q^-, depending on the pH. At low pH values, both the 7- piperazinyl group and 3-carboxyl group are protonated, whereas at high pH values, neither is protonated. The carboxyl group is normally a stronger acid than the ammonium group, the reason being that the neutral nonionic form is spontaneously rearranged to the zwitter ion. The pK_{a1} of CF was found at around 6 and the pK_{a2}, which is due to the presence of an ionizable proton on the external piperazinyl nitrogen, was found at around 8.5. As in our experimental condition, upon heating, chloroauric acid liberates hydrogen chloride, giving gold(III) chloride. In aqueous solution, chloroauric acid consists of the square planar $AuCl_4^-$ ion which is a common precursor to other gold coordination complexes [23]. The possible reaction mechanism in the formation of ciprofloxacin capped gold nanoparticles is shown in Equations 1-3. Tom et al have used CF molecules as a capping agent for Au-NPs in their study. Further, it was proved that piperazinyl ring is modified by the adsorption on gold nanoparticles, possibly through nitrogen whereas pyridone moiety is unaffected [15].

Next, synthesized Au-NPs were re-dispersed in de-ionized water by sonication. 10 μL of the Au-NPs dispersion was casted onto a pre-cleaned GCE surface and the electrode was dried in an air oven at 60°C. Thereafter, Au-NPs modified GCE was used to record cyclic voltammograms (CVs) in 0.1 M KCl containing 5 mM $[Fe(CN)_6]^{3-/4-}$ for 10 cycles at a scan rate of 20 mVs^{-1} (Figure 3, curve a). Further, CVs of a bare GCE in the same condition (Figure 3 curve b) were compared with the CVs of Au-NPs modified GCE which was shown enhanced redox peak currents (curve a). In addition, peak-to-peak separation for $[Fe(CN)_6]^{3-/4-}$ redox peak was 0.09 V which is lower than observed at a bare GCE (0.20 V). This observation elucidated that Au-NPs were attached onto the electrode surface and increased the real surface area. The area of the unmodified and Au-NPs modified electrodes were calculated to be 0.0707 and 0.1139 cm^2, respectively. To find out the fouling of the electrode, CVs were recorded in 0.1 M KCl using a Au-NPs modified electrode which was previously used for measurement of CVs in $[Fe(CN)_6]^{3-/4-}$ + 0.1 M KCl. The Au-NPs modified electrode did not shown any signal for adsorbed $[Fe(CN)_6]^{3-/4-}$ in blank electrolyte, its confirmed that negatively charged redox probe molecules are not retained on the modified electrode surface (Figure 3 curve c).

$$2HAuCl_4 \rightleftharpoons Au_2Cl_6 + 2HCl \quad (1)$$

$$AuCl_4^- + 2HQ \xrightarrow{pH<6} [Au(HQ)_2Cl_4]^- \quad (2)$$

$$[Au(HQ)_2Cl_4] + 4NaOH \xrightarrow{pH>6} Au^0 + 4NaCl + 2HQ^+ + 2H_2O + 2OH^- \quad (3)$$

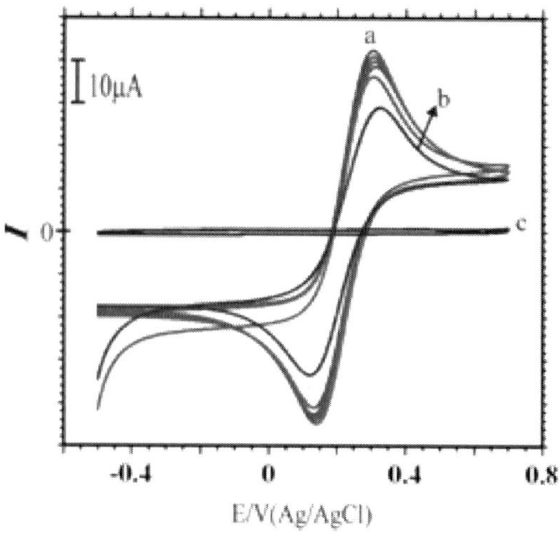

Figure 3. CVs recorded for 10 cycles at Au-NPs modified electrode in 5 mM $[Fe(CN)_6]^{3-/4-}$ + 0.1 M KCl (curve a). CVs recorded at bare-GCE in the same condition (curve b). After running continues CVs for 10 cycles in 5 mM $[Fe(CN)_6]^{3-/4-}$ + 0.1 M KCl, the Au-NPs/GCE was transferred into blank 0.1 M KCl, and CVs were recorded in the same potential window (curve c). Scan rate= 20 mVs^{-1}.

Field-emission SEM measurements were made to investigate the surface morphology and shape of the Au-NPs. Figure 4a shows the representative SEM image obtained for the nanoparticles synthesized by CF. It vouched that nanoparticles with spherical structure were formed. The average size of the nanoparticles was calculated to be 20 nm. Brias et al. reported that pH value of solution influenced on the size of the nanoparticles during the synthesis of Au-NPs by reduced glutathione (GSH). Au NPs coated with GSH were synthesized at various pH values and they found that higher pH decreased sizes from 6 to 2 nm [24]. In the present study, Au-NPs were synthesized with an average size of 20 nm using CF by adjusting pH of the precursors. As can be seen in Figure 4a, the synthesized nanoparticles do not cluster together, likely due to the presence of CF molecules attached to their surface. Obviously the CF's not only stabilize the nanoparticles when suspended in water but prevent them from aggregating when dried in air as well. The CF attached to the nanoparticles surface is visible in the SEM images (Figure 4a).The as-synthesized Au-NPs were characterized using EDS analyses. Figure 4b shows an EDS spectrum obtained at the selected area in SEM image of Au-NPs (Figure 4a). Figure 4b shows that the peaks for Au dominate the spectrum although there are considerable amounts of C, Si and In. Au-NPs were coated on indium tin oxide coated glass for investigation of EDS analysis which is the reason for other peaks. Thus, it is confirmed that the desired element Au-NPs has been effectively synthesized by CF. Further, surface morphology and shape of the nanoparticles were investigated by high-resolution transmission electron microscope (TEM). Figure 4c and d shows the TEM images obtained for the Au-NPs synthesized by CF. Interestingly, Figure 4c and d shows an average spherical particle to be around 20 nm, which is in clear agreement with SEM data.

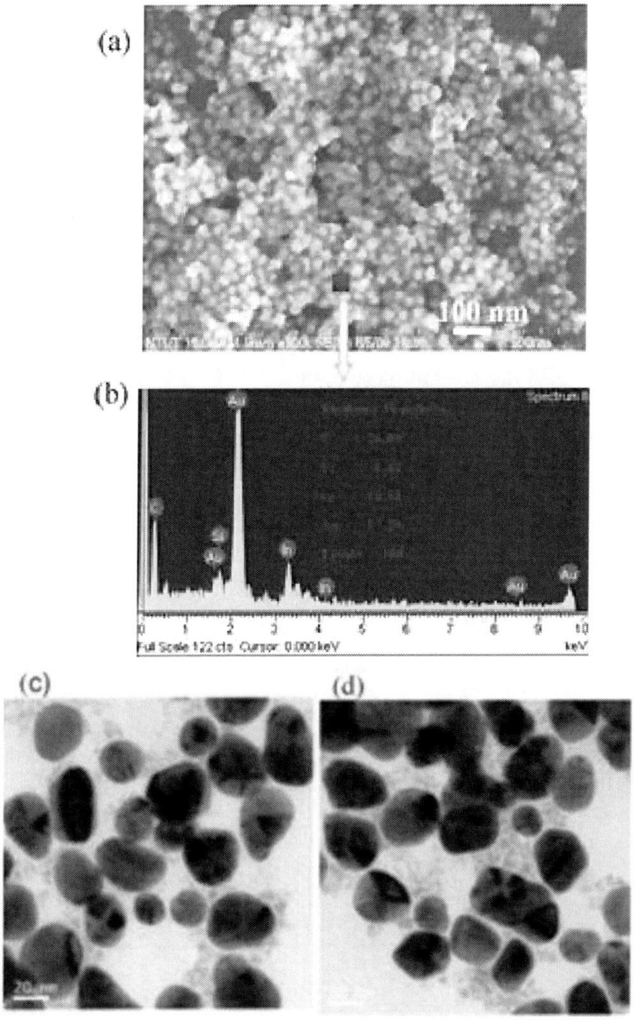

Figure 4. (a) SEM image of Au-NPs coated on ITO electrode and (b) Corresponding EDS spectra of Au-NPs. (c and d) TEM images of synthesized Au-NPs.

As can be seen in Figure 5a the diffraction patterns for the Au-NPs coated ITO show several Bragg-like features than bare surface (Figure 5b). Their position and relative intensity match those of the sharp Bragg peaks observed in the diffraction pattern of bulk crystalline gold [25]. Figure 5a is the XRD pattern of the Au-NPs which reveals that the nanoparticles have fcc structure corresponding to (111), (200), (220), and (311) gold crystalline facets [25].

Figure 6 (curve a) shows cyclic voltammetric data of Au-NPs modified GCE in 0.1 M NaOH. A broad oxidation current flow was evident at +0.38 V on the positive sweep which was accompanied by a large and sharp reduction wave at 0.15 V on the reverse sweep. These two waves are attributed to the formation and reduction of surface Au oxide (AuOx) or Au(OH)x on the nanoparticles [13, 26, 27]. Figure 6 shows two typical sets of cyclic voltammetric data obtained using an Au-NPs/GCE and unmodified GCE electrodes in alkaline solution containing methanol. A large oxidation peak was observed at +0.34 V for methanol oxidation on Au-NPs modified electrode (Figure 6 curve b). This kind of catalytic effect for methanol was not observed at the unmodified GCE in the potential window used, indicating the Au-NPs were involved in the catalytic oxidation of methanol (curve c). In the absence of methanol, CVs responsible for formation and reduction of surface Au oxide were retained as shown Figure 6 curve a.

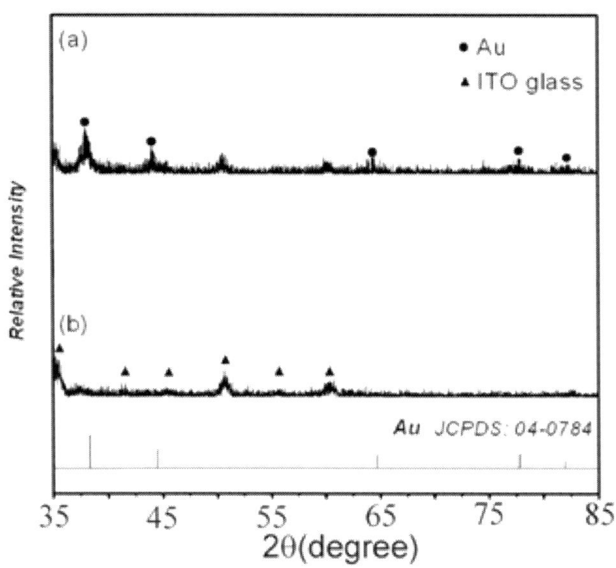

Figure 5. XRD pattern of (a) Au-NPs modified ITO and (b) bare-ITO.

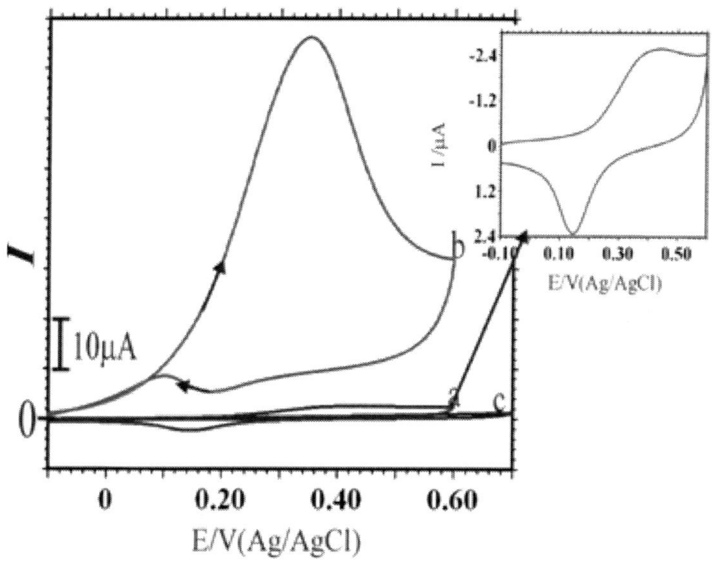

Figure 6. Cyclic voltammetric responses recorded at Au-NPs modified electrode in the absence (curve a) and presence of 800 mM methanol (curve b) in 0.1 M NaOH. The oxidation of methanol at an unmodified electrode in the same conditions (curve c). Scan rate=20 mVs^{-1}.

The voltammetric response observed in the presence of methanol is solely due to its oxidation by the Au-NPs attached GCE. As early observed, the reduction peak observed at + 0.15 V for Au-NPs was disappeared in the presence of methanol in alkaline solution, suggesting that electrogenerated Au-oxide species are involved in the oxidation of methanol [13]. It's well documented that the surface oxides of Au-NPs can function as an electron-transfer mediator in the oxidation process of methanol [26, 27].

Figure 7A shows the voltammetric dependence on scan rate (ν) for Au-NPs in 0.1 M NaOH containing methanol. Plot of oxidation current of methanol vs. scan rate was linear from 20 to 200 mVs^{-1}, indicating surface-controlled process for methanol oxidation (Inset of Figure 7A). All these results show that the oxidation of methanol is mediated by surface oxide redox species. Further to support it, voltammetric responses were recorded for different concentrations of methanol using an Au-NPs coated GCE (Figure 7B). The gradual decrease in the cathodic peak corresponding to the reduction of surface oxide on increasing the concentration of methanol in solution demonstrates the involvement of surface oxides in the catalytic reaction (Figure 7B).

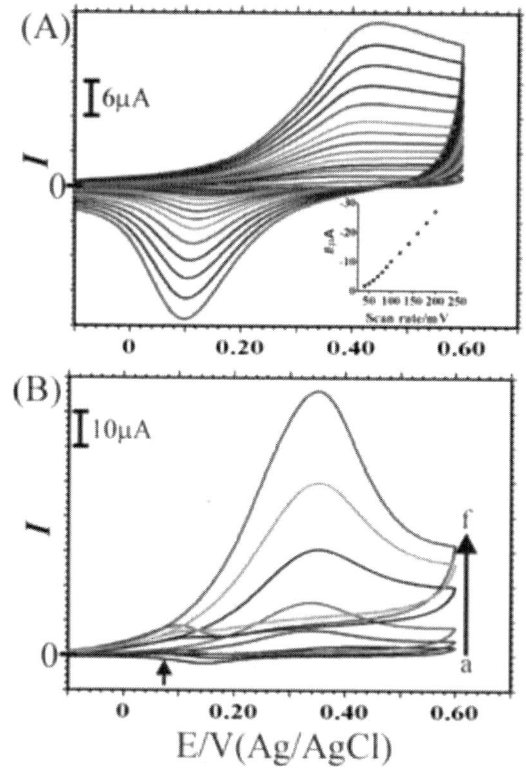

Figure 7. (A) CVs for the oxidation of methanol (82 mM) at Au-NPs/GCE in 0.1 M NaOH. Scan rate: (a) 20, (b) 40, (c) 50, (d) 60, (e) 70, (f) 80, (g) 90, (h) 100, (i) 120, (j) 140, (k) 160, (l) 180 and (m) 200 mVs^{-1}. Inset shows corresponding plot of peak current against scan rate. (B) CVs for the oxidation of methanol at Au-NPs/GCE in 0.1 M NaOH at different concentration. Each addition increased the concentration of methanol by 160 mM. Scan rate=20 mVs^{-1}.

Figure 8 show the cyclic voltammetric response of Au-NPs modified GCE in O_2-free (curve a) and O_2-saturated (curve b) phosphate buffer solution. In the presence of O_2 a significant increase in the cathodic peak at about −0.33 V was observed, indicating electrocatalytic effect of attached Au-NPs. At an unmodified GCE such characteristic reduction wave was not observed in O_2-saturated buffer solution in the potential range used (Figure 8 curve c). The reduction peak observed at -0.33 V corresponds to the reduction of oxygen to H_2O.

The mechanism for reduction of oxygen at Au-NPs modified electrode in pH 7.2 buffer solution was believed to follow a 4-electron reduction of O_2 to H_2O (or OH^-) [28, 29].

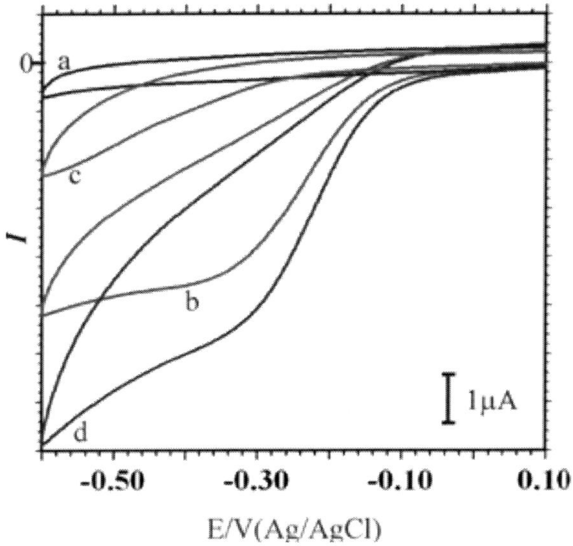

Figure 8. Electrocatalytic reduction of oxygen at Au-NPs modified electrode (b) and unmodified electrode (c). CVs of Au-NPs modified electrode in O_2-free buffer solution (curve a, pH 7.2). Cyclic voltametric response recorded at Au-NPs modified electrode for addition of 1.8×10^{-3} M H_2O_2 in O_2-saturated solution (curve d). Scan rate= 20 mVs^{-1}.

To support this conclusion, a known amount of H_2O_2 (1.8×10^{-3} M H_2O_2) was added into O_2-saturated buffer solution and the cyclic voltammetric response was recorded. It was found that cathodic current was increased at -0.33 V (Figure 8 curve d), confirming that final product of O_2 reduction was H_2O. It was further supported by Jena and Raj, who observed that spherical gold nanoparticles showed a single reduction peak for catalytic reduction of oxygen and H_2O_2 [13]. In addition, Au-NPs modified GCE showed a cathodic peak of O_2 reduction at same potential (-0.33 V) suggesting that spherical Au-NPs were successfully synthesized by CF, and they have good catalytic effect towards reduction of oxygen.

CONCLUSION

Selecting the right reducing agent for NPs synthesis is very important in forming Au-NPs with desirable sizes. In this study, we had chosen the synthetic antibiotic ciprofloxacin as a reducing/capping agent. In this chapter, we reported synthesis of water-soluble gold nanoparticles using CF. Because of the favorable properties such as the presence of -C=O, carboxylic acid, and imino groups, water solubility at relevant biological pH and biological compatibility, CF is a very attractive ligand in making water-soluble Au-NPs for catalysis applications. Spectral and electrocatalytic studies of synthesized Au-NPs were reported. TEM, SEM combined with EDS showed spherical shaped nanoparticles with average size of ~20 nm was formed. Electrocatalytic oxidation of methanol and reduction of oxygen were observed at Au-NPs modified GCE. The Au-NPs modified electrode shows a single voltammetric peaks for a four-electron reduction of oxygen at a less-negative potential in neutral pH, which is not observed at unmodified electrode in the potential range used. We have described a simple method for the preparation of Au-NPs and expected to have major application in the electrocatalysis of methanol and oxygen for energy source applications.

REFERENCES

[1] M. C. Daniel, D. Astruc, *Chem. Rev.* 104 (2004) 293-346.

[2] T. Kim, K. Lee, M. S. Gong, S. W. Joo, *Langmuir* 21 (2005) 9524-9528.

[3] T. Zidki, H. Cohen, D. Meyerstein, *Phys. Chem. Chem. Phys.* 8 (2006) 3552-3556.

[4] D. Prosperi, C. Morasso, F. Mantegazza, M. Buscaglia, L. Hough, T. Bellini, *Small* 2 (2006) 1060-1067.

[5] G. F. Paciotti, L. Myer, D. Weinreich, D. Goia, N. Pavel, R. E. McLaughlin, L. Tamarkin, *Drug Delivery* 11 (2004) 169-183.

[6] S. R. Narayanan, T. I. Valdez, *In Handbook of Fuel Cells: Fundamentals, Technology, and Applications;* Vielstich, W., Gasteiger, H. A., Lamm, A., Eds.; John Wiley and Sons: Chichester, England, 2003; Vol. 4, pp 1133-1141.

[7] L. A. Bauer, N. S. Birenbaum, G. J. Meyer, *J. Mater. Chem.* 14 (2004) 517–526.

[8] D. A. Handley, *In Colloidal Gold: Principles, Methods, and Applications*; Hayat, M. A., Ed.; Academic Press, Inc.: San Diego, 1989; Vol. 1, pp 13-30.

[9] L. L. Rouhana, J. A. Jaber, J. B. Schlenoff, *Langmuir* 23 (2007) 12799-12801.

[10] F. Hubert, F. Testard, O. Spalla, *Langmuir* 24 (2008) 9219-9222.

[11] K. Sun, J. Qiu, J. Liu, Y. Miao, *J. Mater. Sci.* 44 (2009) 754–758.

[12] K. B. Male, J. Li, C. C. Bun, S. C. Ng, J. H. T. Luong, *J. Phys. Chem.* C 112 (2008) 443-451.
[13] B. K. Jena, C. R. Raj, *Langmuir* 23 (2007) 4064-4070.
[14] H. Zhang, J. J. Xu, H. Y. Chen, *J. Phys. Chem.* C 112 (2008) 13886–13892.
[15] R. T. Tom, V. Suryanarayanan, P. G. Reddy, S. Baskaran, T. Pradeep, *Langmuir* 20 (2004) 1909-1914.
[16] L. A. Mitscher, *Chem. Rev.* 105 (2005) 559-592.
[17] K. E. Brighty, T. D. Gootz, in: V.T. Andriole (Ed.), *The Quinolones*, Academic Press, San Diego, 2000, p. 33.
[18] G. Sheehan, N. S. Y. Chew, in: A.R. Ronald, D.E. Low (Eds.), *Fluoroquinolone Antibiotics*, Birkhauser, Basel, Switzerland, 2003, p. 1.
[19] P. Trivedi, D. Vasudevan, *Environ. Sci. Technol.* 41 (2007) 3153–8.
[20] S. A. Kumar, S. F. Wang, *Mat. Let.* 63 (2009) 1830–1833.
[21] J. Qiu, X. Jiang, C. Zhu, M. Shirai, J. Si, N. Jiang, K. Hirao, Angew. *Chem. Int. Ed.* 43 (2004) 2230–2234.
[22] Turel, Coord. *Chem. Rev.* 232 (2002) 27-47.
[23] N. N. Greenwood, A. Earnshaw, (1997). *Chemistry of the Elements* (2nd Edn.), Oxford:Butterworth-Heinemann. Chapter 28, page no. 1173. ISBN 0-7506-3365-4.
[24] R. P. Brinas, M. Hu, L. Qian, E. S. Lymar, J. F. Hainfeld, *J. Am. Chem. Soc.* 130 (2008) 975-982.
[25] V. Petkov, Y. Peng, G. Williams, B. Huang, D. Tomalia, Y. Ren, Phys. Rev. B: 72 (2005) 195402.
[26] L. D. Burke, P. F. Nugent, *Gold Bull.* 30 (1997) 43–53.
[27] L. D. Burke, P. F. Nugent, *Gold Bull.* 31 (1998) 39–50.
[28] C. Paliteiro, N. Martins, *Electrochim. Acta* 44 (1998) 1359-1368.
[29] M. S. El- Deab, T. Ohsaka, *Electrochem. Commun.* 4 (2002) 288–292.

INDEX

B

C

D

E

F

G

H

I

M

N

O

P

V

W

X

Z